Statistical Mechanics
and Spectroscopy

Statistical Mechanics and Spectroscopy

G. ALLEN
Professor of Chemical Physics,
University of Manchester, England

H. O. PRITCHARD
Professor of Chemistry,
York University, Canada

A HALSTED PRESS BOOK

JOHN WILEY & SONS
New York–Toronto

English edition first published in 1974 by Butterworth & Co
(Publishers) Ltd
88 Kingsway, London WC2B 6AB

Published in the U.S.A. and Canada by Halsted Press, a Division of John Wiley & Sons, Inc.,
New York

Library of Congress Cataloging in Publication Data

Allen, Geoffrey.
Statistical mechanics and spectroscopy.

'A Halsted Press book.'
1. Molecular spectra. 2. Statistical mechanics.
I. Pritchard, Huw Owen, 1928–joint author.
II. Title.
QC454.M6A4 1974 530.1'33 74-7034
ISBN 0-470-02331-7

Printed in England by
The Whitefriars Press Ltd, London and Tonbridge.

PREFACE

This book is based on a lecture course given by the authors to second-year undergraduate chemistry students in the University of Manchester. We consider that an elementary knowledge of statistical mechanics and of molecular spectroscopy, among other things, is essential to a proper understanding of physical chemistry. Furthermore, the two topics are so intertwined and interdependent that it is difficult to develop either one of them adequately without a working knowledge of the other. In this course we set out to introduce the students to the basic facts of both topics simultaneously and in a logical manner. To do this, we assume that they already have been exposed to an elementary course of quantum chemistry and are aware of the existence of the Schrödinger equation and, in a qualitative way, some of the properties of its solutions; also that they are aware of the basic facts of experimental chemical thermodynamics. We believe it is inevitable that an exposition at this elementary level will gloss over many difficult points, and we make no assertions whatever as to the rigour of the arguments presented.

Both authors wish to acknowledge their debt to the late Professor M. G. Evans, who not only stimulated their interest in physical chemistry during their student days but also excelled in the teaching of the principles of statistical mechanics.

G.A.
H.O.P.

ACKNOWLEDGEMENTS

We are grateful to the following for permission to reproduce copyright material: Masson et Cie, Paris for the use of Figure 1 from Herzberg, 'Atomic Spectra and Atomic Structure', *Annales de Physique*, Vol. 84, No. 4 (1927); Dover, New York for the use of a figure from Herzberg, *Atomic Spectra and Atomic Structure*; Cambridge University Press for the use of two figures from Condon and Shortley, *The Theory of Atomic Spectra*; McGraw-Hill, New York for the use of Figures 2.4 and 10.7 from Barrow, *Introduction to Molecular Spectroscopy*, and Figure 8.3 from Davidson, *Statistical Mechanics*; Van Nostrand, London for the use of Figure 9.15 from Sugden and Kenny, *Microwave Spectroscopy of Gases*; Chapman and Hall, London for the use of Figure 2.4 from Gordy, Smith and Trambarulo, *Microwave Spectroscopy*; McGraw-Hill (U.K.) Ltd. for the use of Figures 3.9, 3.12, 3.14, 3.17 and 3.18 from Banwell, *Fundamentals of Molecular Spectroscopy*; Wiley, New York for the use of Figure 4 from Hill and Day, *Physical Methods in Advanced Inorganic Chemistry*; the Clarendon Press for the use of Figures 12, 13, and 15 from Rushbrooke, *Introduction to Statistical Mechanics*; the Academic Press for the use of Figure 2.6 from Gordy, Smith and Trambarulo, *Microwave Spectroscopy*, reproduced from an original diagram by Coles, *Advances in Electronics*, Vol. 11 (1950).

Also, we thank: Van Nostrand, New York for the use of figures from Herzberg, *Spectra of Diatomic Molecules* of which Figure 3 was reproduced from an original diagram by Burmeister, *Bericht der Deutschen Physikalischen Gesellschaft*, Vol. 15 (1913); Springer, Berlin for the use of figures from Grotrian, *Graphische Darstellung der Spektren von Atomen und Ionen mit ein, zwei und drei Valenzelektronen*; the American Institute of Physics for the use of the following material from *Physical Review*: a figure by Standberg, Wentink and Kyhl, Vol. 75 (1949); one figure from Daking, Good and Coles, Vol. 70 (1946); a figure from Good, Vol. 70 (1946); a figure from Simmons, Anderson and Gordy, Vol. 77 (1950); a figure by Brice, Vol. 38 (1931) after which Figure 6.10(a) was drawn; and a figure from Hopfield and Birge, Vol. 29 (1927) after which Figure 6.10(b) was drawn; Dr. J. A. Marquisee for the use of his Figures 10.3, 10.4 and 10.6 which were published by McGraw-Hill in Barrow, *Introduction to Molecular Spectroscopy*; the Berlin Physikalische Gesellschaft for the use of a figure from Schuler, Hahn and Gallnow, *Zeitschrift fur Physik,* Vol. 111 (1939).

CONTENTS

PUBLISHER'S NOTE

We should like to draw the reader's attention to the use of the symbols υ (denoting frequency) and ν (denoting quantum number of a vibrational state). Unfortunately, in the preparation of this book it was not possible to match the symbols used for these quantities in the figures with those used in the text. In the figures, frequency is indicated by ν and the quantum number of a vibrational state by v, while in the text these quantities are υ and ν, respectively. We apologise for this dichotomy and hope that it will not prove too confusing.

Chapter One

AN INTRODUCTION TO STATISTICAL MECHANICS

The Nature of the Problem

The properties of individual molecules can be understood, in principle at least, in terms of quantum mechanics. For example, we can calculate with great accuracy the properties of small atoms such as H, He, or small molecules such as H_2^+ or H_2, and we are confident that we could do likewise for more complicated systems if we could solve the relevant Schrödinger equation. However, the solution of the Schrödinger equation tells us only the properties of a *single system,* e.g. a single molecule, or an atom, or an ion. It tells us nothing about the properties of an *assembly* of such systems, e.g. 1 litre of hydrogen gas. This is in the realm of classical thermodynamics, and it is the aim of statistical thermodynamics (or statistical mechanics) to bridge this gap between the quantum-mechanical description of single molecules and the bulk properties of the substance as we encounter it in the laboratory. The foundations of this subject were laid towards the end of the last century by Maxwell and Boltzmann, and consolidated in the early decades of this century by Gibbs, Darwin, Fowler and others.

It is very simple to show that some sort of relationship exists between quantum mechanics and thermodynamics. If we have N non-interacting atoms, each of mass m, confined in a volume of dimensions $a \times b \times c = V$, we can treat the motion of each atom as a 'particle in a box'. Thus the component of energy along the a-axis is

$$\epsilon_i^{(a)} = \frac{i^2 h^2}{8ma^2} \qquad (1.1)$$

where the quantum number i can take the values 1, 2, 3, . . ., etc., and h is Planck's constant. There are similar components along the b- and c-axes. If there are n_i atoms in each quantised energy level, then the total a-component of the energy of the assembly is

$$E^{(a)} = n_1 \epsilon_1^{(a)} + n_2 \epsilon_2^{(a)} + \ldots n_i \epsilon_i^{(a)} + \ldots \tag{1.2}$$

where

$$\sum_i n_i = N \tag{1.3}$$

The energy of the assembly will change if we do anything which alters the absolute values of the ϵ_i or the distribution of the numbers n_1, n_2, . . ., n_i, . . ., etc. For example, if we double the total number of atoms to $2N$ without altering the distribution, we double the total energy, but the average energy per atom remains unchanged. Alternatively, if we change the volume by altering the length a, all the values of the $\epsilon_i^{(a)}$ change in equation 1.1 and so the total energy changes. Finally, if we make some change which alters the distribution numbers n_i, then the energy changes according to equation 1.2. We shall see that these changes are effectively changes in P, V and T, respectively. Unfortunately, however, this kind of approach does not give us the quantitative correspondence between the properties of the individual systems and the thermodynamic properties of the assembly for which we are looking, because we have no criterion by which to determine the distribution numbers n_i. To do this, we have to consider the much more subtle concept of entropy.

Entropy and Randomness

In classical thermodynamics we distinguish between two types of process: one we call reversible; the other, irreversible. We then define a quantity called entropy, S, which has the property that it remains unchanged during a reversible process. However, any spontaneous or irreversible change is accompanied by an increase in entropy. The whole key to statistical mechanics is to find some property associated with an assembly of molecules which simulates this behaviour. A little thought will show that the mechanical concept of *disorder* or randomness has the desired properties, and it is in fact the link for which we are looking.

For example, if we stack the cards after a hand of bridge, they are more or less arranged in groups of four of the same suit. But when we shuffle the pack, this ordering is destroyed and the distribution of cards becomes more random. Again, if a poker-dice player is passed four aces and a queen, he will gain little by shaking all five dice! In both cases the natural tendency is towards an increase in disorder. At a molecular level we can see that disorder has properties which run parallel to those observed for entropy.

For example, if two different gases are separated by a barrier, the total entropy increases when we remove that barrier and allow the gases to mix in a spontaneous, irreversible process. At the same time, a certain amount of order which existed in the distribution of the molecules is destroyed, as in the pack of cards we have just shuffled.

We are going to identify this concept of randomness with the number of *a priori equally probable arrangements* of the systems (e.g. atoms, etc.) in the assembly, and we will denote this quantity by Ω. But as we do not yet know which molecular properties become more randomly distributed as entropy increases, let us progress a little further with an example which does not refer to molecules.

We have a pack of cards which we shuffle continuously. After a long time, any of the possible arrangements of the cards is equally probable, and there are Ω such arrangements. Now suppose that we have two *separate* packs with Ω_1 and Ω_2 equally probable arrangements (Ω_1 will equal Ω_2 if the two packs contain equal numbers of cards). The total number of equally probable arrangements of both packs is $\Omega = \Omega_1 \times \Omega_2$, since for each possible arrangement of pack 1 there are Ω_2 possible arrangements of pack 2. Returning to our thermodynamic experience, we know that, if we have two *independent* assemblies having entropies S_1 and S_2, the total entropy is $S = S_1 + S_2$. We note that probabilities are multiplicative, whereas entropies are additive. We therefore postulate, following Planck, that there is a correspondence between the two concepts, involving a constant k, of the form

$$\begin{aligned} S &= k \ln \Omega \\ &= k \ln \Omega_1 \Omega_2 = k \ln \Omega_1 + k \ln \Omega_2 \\ &= S_1 + S_2 \end{aligned} \tag{1.4}$$

and, since experimentally we know that S is a function of E (the total energy), V (the volume) and N (the number of systems in the assembly), i.e.

$$T\,\mathrm{d}S = \mathrm{d}E + P\,\mathrm{d}V - \mu\,\mathrm{d}N$$

we likewise postulate that Ω is a function of these same three quantities. Consequently we can write

$$\left(\frac{\partial S}{\partial E}\right)_{V,N} = \frac{1}{T} = k\left(\frac{\partial \ln \Omega}{\partial E}\right)_{V,N} \tag{1.5}$$

$$\left(\frac{\partial S}{\partial V}\right)_{E,N} = \frac{P}{T} = k\left(\frac{\partial \ln \Omega}{\partial V}\right)_{E,N} \tag{1.6}$$

$$\left(\frac{\partial S}{\partial N}\right)_{E,V} = -\frac{\mu}{T} = k\left(\frac{\partial \ln \Omega}{\partial N}\right)_{E,V} \tag{1.7}$$

and this brings us to the crux of the whole problem. If we can calculate Ω as a function of E, V and N, then we can derive the thermodynamic properties, temperature, pressure and chemical potential (T, P, μ) and of course S—in other words, virtually a complete thermodynamic description of the assembly.

But we have said we do not know which property it is whose randomness is important. For instance, it could be position, velocity, momentum, kinetic energy, potential energy, and so on. And here we come to a vital step in the argument, which we cannot prove. We do not know which of these properties to choose, but our experimental knowledge of the properties of S suggests that it may be the total energy E, which is the sum of the energies of the individual systems. Proceeding on this basis, we shall eventually see that the postulate is justified on the basis of the results we obtain. It is interesting to note here that both statistical mechanics and quantum mechanics depend on certain axiomatic foundations—that is, in both cases there are some fundamental equations which we cannot rigorously derive. In quantum mechanics we cannot rigorously derive the Schrödinger equation. There are many ways in which the equation can be obtained, but they are all based in one way or another on the analogy between particle and wave properties embodied in the de Broglie relationship. Nor can we derive the Pauli exclusion principle, although its acceptance is vital to the quantum-mechanical description of atoms and molecules. Likewise, in statistical mechanics we start with the analogy between randomness and entropy, and we justify the further postulates we need because they are the only ones which lead to the correct results.

The Boltzmann Distribution

We consider an assembly* of N systems* with a total energy E, contained in a volume V. The motion of each system is quantised, whether it be translational motion or any of the possible motions of one part of the system with respect to another, i.e. electronic, vibrational or rotational motion. Associated with each type of motion, there are only certain allowed energy levels, which usually fall into a fairly simple pattern, as we shall see later. Thus the total energy ϵ, i.e. the sum of all the individual components, can only take certain values, although the pattern of allowed energy levels for each system is a rather complicated one. These allowed energy levels, irrespective of the way in which the total energy is built up, are labelled ϵ_1, ϵ_2, ϵ_3, . . ., etc., in ascending order of energy.

One of our fundamental postulates is that it is this total energy E to which the concept of randomness applies. But before we can go any further, we need to make a second fundamental postulate: that *a priori* each allowed energy level of the assembly has an equal probability, in the same way as any distribution of cards in a pack is equally probable. This concept of equal *a priori* probabilities is complicated somewhat by the phenomenon of degeneracy in quantum mechanics. For example, there are four solutions of the Schrödinger equation for the hydrogen atom with a principal quantum number $n = 2$, i.e. one 2s and three 2p functions, all of which correspond to the same energy ϵ_2. Similarly, for $n = 3$, there are nine degenerate solutions having energy ϵ_3. Each of these degenerate solutions is regarded as distinct, so that the *a priori* probabilities of the states of the hydrogen atom with $n = 1, 2, 3, \ldots$ are taken to be 1, 4, 9, . . ., respectively.

We imagine our assembly to consist of N identical systems, n_1 of which are in energy level ϵ_1, n_2 in ϵ_2, . . ., n_i in ϵ_i, and so on, so that

$$\sum_i n_i = N \tag{1.8}$$

$$\sum_i n_i \epsilon_i = E \tag{1.9}$$

and we wish to count up the number of possible ways in which such a distribution can be achieved if the N systems are shuffled among the energy levels. We will consider the energy levels as boxes which are identifiable only by their energy ϵ_i, and we will assign to each a

* Recall from p. 1 the definitions of these terms.

corresponding degeneracy g_i. Thus we have a very complicated arrangement of energy levels:

$$\epsilon_1\ \epsilon_2\ \epsilon_3, \ldots, \epsilon_i, \ldots,$$

each of degeneracy

$$g_1\ g_2\ g_3, \ldots, g_i, \ldots,$$

and each containing

$$n_1\ n_2\ n_3, \ldots, n_i, \ldots,$$

systems.

For a moment, let us ignore the degeneracy. We are interested in the number of ways in which we can arrange the individual systems, to give the defined assembly, whose properties are partially determined by equations 1.8 and 1.9. For each of these arrangements, if we take a pair of systems and interchange their properties, the properties of the assembly remain unchanged. We have N systems, so the total number of possible arrangements is $N!$. However, not all these permutations give rise to different states of the assembly. We will identify the quantity Ω with those which are significant perturbations, i.e. which give different states of the assembly, and write

$$N! = \Omega \times \text{(number of insignificant permutations)} \qquad (1.10)$$

We mean by a significant perturbation: taking one of the n_1 systems out of box ϵ_1 and exchanging it for one of the n_3 in box ϵ_3, say. On the other hand, to change around two of the n_1 systems in box ϵ_1 is not regarded as significant.* The number of ways we can interchange pairs among n_1 systems is $n_1!$, and likewise $n_2!$ for n_2, $n_i!$ for n_i, etc. Thus, equation 1.10 becomes

$$N! = \Omega n_1!\, n_2!\, n_3! \ldots$$

or

$$\Omega = \frac{N!}{n_1!\, n_2!\, n_3! \ldots} \qquad (1.11)$$

* A simple analogy would be to suppose that we have five dice arranged in a straight line showing A A A K K: if we interchange the properties (in this case, position) of the first two, we still have the same arrangement or complexion A A A K K; however, if we interchange the first and last, we have a new complexion K A A K A.

Let us now consider degeneracy. We have to remember that the box we labelled ϵ_1 is not really one box but g_1 boxes. The number of ways we can put n objects into g boxes is g^n (we can put the first object in g ways, the second one in g ways, i.e. g^2 ways altogether; the third one in g ways, i.e. g^3 ways altogether, and so on up to n objects).† So equation 1.11 should really be

$$\Omega = \frac{N!\, g_1^{n_1}\, g_2^{n_2}\, g_3^{n_3} \ldots}{n_1!\, n_2!\, n_3! \ldots} = \frac{N! \prod_i g_i^{n_i}}{\prod_i n_i!} \tag{1.12}$$

Since we are seeking the most probable state of the assembly, we have now to choose our distribution $n_1, n_2, n_3, \ldots$ so that the entropy S, and therefore $k \ln \Omega$ (see equation 1.4), is a maximum, i.e. we must maximise

$$\ln \Omega = \ln N! - \sum_i \ln n_i! + \sum_i n_i \ln g_i \tag{1.13}$$

Before we can proceed, we have to learn how to handle awkward quantities like $\ln N!$, which we do by using Stirling's approximation for $N!$, i.e.

$$N! \simeq (2\pi N)^{1/2} N^N e^{-N} \tag{1.14}$$

Table 1.1 shows the accuracy of Stirling's approximation. Remembering that N is of the order of 10^{23}, we might expect an error of the order

Table 1.1 ACCURACY OF STIRLING'S APPROXIMATION

N	$N!$	*Eq. 1.14*	*Error*
1	1	0.922	8%
2	2	1.919	4%
5	120	118.019	2%
10	3.628×10^6	3.598×10^6	0.8%
100	9.330×10^{157}	9.322×10^{157}	0.08%

of 8×10^{-23}%, so that the use of this approximation does not give rise to any meaningful error. Taking logarithms of equation 1.14, we have

$$\ln N! \simeq (N + \tfrac{1}{2}) \ln N - N + \tfrac{1}{2} \ln 2\pi \tag{1.15}$$

† To continue with our dice analogy: each face has a degeneracy of 4 in the sense that it may be oriented A, ᗒ, ∀ or ᗕ. Thus there are 4^3 distinguishable complexions of the type AAA, A∀A, ᗒᗒᗒ, etc. if three dice are used.

but for such large values of N, we can safely neglect both the $\frac{1}{2}$ in the first bracket and the $\frac{1}{2}$ ln 2π term, obtaining

$$\ln N! \simeq N \ln N - N \tag{1.16}$$

Since we are looking for a maximum, we must differentiate quantities of the sort $\ln N!$, say with respect to N. Equation 1.16 enables this to be done:

$$\frac{\mathrm{d} \ln N!}{\mathrm{d}N} = \frac{\mathrm{d}(N \ln N - N)}{\mathrm{d}N} = (\ln N + 1 - 1) = \ln N \tag{1.17}$$

In order to find the distribution $n_1, n_2, n_3, \ldots$ which maximises $\ln \Omega$, we differentiate $\ln \Omega$ with respect to each of these occupation numbers in turn. However, it is not quite as simple as this because we cannot vary the n_i values indiscriminately; we can only change them in such a way that the two restrictive conditions of equations 1.8 and 1.9 are obeyed, since we do not want to change the total number of systems in the assembly, or the total energy of the assembly. Thus we have to be rather careful about how we carry out the differentiation. We make a series of small changes $\delta n_1, \delta n_2, \delta n_3, \ldots$ in $n_1, n_2, n_3, \ldots$, respectively, and correspondingly there is a small change $\delta \ln \Omega$ in $\ln \Omega$ given by

$$\delta \ln \Omega = \frac{\partial \ln \Omega}{\partial n_1} \delta n_1 + \frac{\partial \ln \Omega}{\partial n_2} \delta n_2 + \frac{\partial \ln \Omega}{\partial n_3} \delta n_3 + \\ + \ldots \frac{\partial \ln \Omega}{\partial n_i} \delta n_i + \ldots \tag{1.18}$$

To maximise $\ln \Omega$, this expression for $\delta \ln \Omega$ must equal zero. At the same time, to satisfy equations 1.8 and 1.9,

$$\delta n_1 + \delta n_2 + \delta n_3 + \ldots \delta n_i + \ldots = 0 \tag{1.19}$$

and

$$\epsilon_1 \delta n_1 + \epsilon_2 \delta n_2 + \epsilon_3 \delta n_3 + \ldots \epsilon_i \delta n_i + \ldots = 0 \tag{1.20}$$

so that N and E remain unchanged. Thus we have three equations which must be satisfied simultaneously; we can combine them into one equation by multiplying each by a different numerical constant and adding the resulting equations. Multiplying equation 1.18 by one, equation 1.19 by α and equation 1.20 by β, we get

$$\delta \ln \Omega = 0 = \left[\frac{\partial \ln \Omega}{\partial n_1} + \alpha + \beta \epsilon_1\right] \delta n_1 + \left[\frac{\partial \ln \Omega}{\partial n_2} + \alpha + \beta \epsilon_2\right] \delta n_2 \ldots +$$

$$+ \left[\frac{\partial \ln \Omega}{\partial n_i} + \alpha + \beta \epsilon_i\right] \delta n_i \ldots \qquad (1.21)$$

A series like this can only be identically zero for arbitrary increments $\delta n_1, \delta n_2$, etc., if all the terms in the brackets are themselves zero, i.e., in general,

$$\frac{\partial \ln \Omega}{\partial n_i} + \alpha + \beta \epsilon_i = 0 \; (i = 1, 2, 3, \ldots) \qquad (1.22)$$

From equations 1.13 and 1.17 we obtain

$$\frac{\partial \ln \Omega}{\partial n_i} = 0 - \ln n_i + \ln g_i \qquad (1.23)$$

Substituting this into equation 1.22 gives

$$\ln n_i = \ln g_i + \alpha + \beta \epsilon_i \qquad (1.24)$$

or, in terms of exponentials,

$$n_i = g_i \, e^{\alpha} \, e^{\beta \epsilon_i} \qquad (1.25)$$

These values of n_i now correspond to our assembly of maximum entropy, and it only remains to find the appropriate values of α and β, which we do by comparison with experiment. With the use of equations 1.13 and 1.17, we write

$$\begin{aligned} S &= k \ln \Omega \\ &= k \left[\ln N! - \sum_i \ln n_i! + \sum_i n_i \ln g_i\right] \\ &= k \left[N \ln N - N - \sum_i (n_i \ln n_i - n_i) + \sum_i n_i \ln g_i\right] \qquad (1.26) \end{aligned}$$

N cancels with $\sum_i n_i$, so that

$$S = k\left[N \ln N - \sum_i n_i(\ln n_i - \ln g_i)\right]$$

Substitution of equation 1.24 leads to

$$\begin{aligned} S &= k\left[N \ln N - \sum_i n_i(\alpha + \beta\epsilon_i)\right] \\ &= k\left[N \ln N - \alpha N - \beta E\right] \qquad (1.27) \end{aligned}$$

Now k is a fundamental constant, N is a constant of the assembly, and α and β are arbitrary constants which were introduced in our derivation. Also, we have taken the volume V to be fixed throughout, so that from equation 1.27 we deduce that

$$\left(\frac{\partial S}{\partial E}\right)_{N,V} = -k\beta$$

which, by comparison with equation 1.5, i.e.

$$\left(\frac{\partial S}{\partial E}\right)_{N,V} = \frac{1}{T}$$

gives

$$\beta = -\frac{1}{kT} \tag{1.28}$$

Thus, the occupation numbers of the distribution of maximum entropy, i.e. the equilibrium distribution, are given by

$$n_i = g_i\, e^{\alpha}\, e^{-\epsilon_i/kT} \tag{1.29}$$

But

$$N = \sum_i n_i = \sum_i g_i\, e^{\alpha}\, e^{-\epsilon_i/kT} = e^{\alpha} \sum_i g_i\, e^{-\epsilon_i/kT} \tag{1.30}$$

and hence

$$e^{\alpha} = N/\sum_i g_i\, e^{-\epsilon_i/kT} \tag{1.31}$$

We shall not be interested in α (but it can be shown that $\alpha = \mu/kT$) and so we eliminate it from equation 1.29 by means of equation 1.31, obtaining finally

$$n_i = \frac{N g_i\, e^{-\epsilon_i/kT}}{\sum_i g_i\, e^{-\epsilon_i/kT}} \tag{1.32}$$

This is the Boltzmann distribution, and the constant k is known as the Boltzmann constant, having a numerical magnitude of $k = 1.380\ 54 \times 10^{-16}$ erg K^{-1} molecule^{-1}; if we are talking in chemical terms and express our energies in cal mol^{-1} or J mol^{-1} instead of erg molecule^{-1}, we simply replace $\exp(-\epsilon_i/kT)$ by the equivalent expression $\exp(-E_i/RT)$, in which $E_i = N\epsilon_i$ and $R = Nk = 1.987$ cal K^{-1} mol^{-1} = 8.314 J K^{-1} mol^{-1}.

To recapitulate, the Boltzmann distribution arises directly from the postulate that entropy is an additive property whereas randomness is multiplicative, and also from the experimental knowledge that entropy is maximised in an equilibrium situation. We have needed, of course, to make some assumptions concerning the concept of randomness in order to calculate Ω, and these we justify on the grounds that they lead to experimentally verifiable results.

The Partition Function

The denominator of equation 1.32 consists of the summation

$$\sum_i g_i \exp(-\epsilon_i/kT)$$

which fulfils the same function in this context as does the normalising integral $\int \psi \bar{\psi} \mathrm{d}\tau$ in quantum mechanics. The quantity $g_i \exp(-\epsilon_i/kT)$ in the numerator of equation 1.32 is the probability of finding the system in energy level i. These probabilities, as they stand, are only correct in a relative sense, i.e. they apply to the comparison of one energy level with another. Also, they will be correct in an absolute sense if they are scaled so that their sum is unity. Hence, we divide by

$$\sum_i g_i \exp(-\epsilon_i/kT)$$

in statistical mechanics. The latter quantity was originally known in German as *Zustandsumme,* i.e. 'state-sum', which is an appropriate term. Unfortunately, with time, the state-sum has become known as the 'partition function'. This name arises because the quantity

$$\sum_i g_i \exp(-\epsilon_i/kT)$$

is supposed to indicate how, at equilibrium, the systems are partitioned among the available energy states, although it is really the individual terms in the summation which give us this information. Beyond this, it is difficult to attach any precise physical interpretation to the partition function, which we shall denote by the symbol f. Nevertheless, it is a very useful quantity and occupies a central position in statistical mechanics, as we shall see below.

Taking logarithms of equation 1.31, we obtain

$$\alpha = \ln N - \ln \sum_i g_i \mathrm{e}^{-\epsilon_i/kT} = \ln N - \ln f \qquad (1.33)$$

Thus, substituting for α in equation 1.27, we have

$$\begin{aligned} S &= k(N \ln N - N \ln N + N \ln f + E/kT) \\ &= Nk \ln f + E/T \qquad (1.34) \end{aligned}$$

Now, the Helmholtz free energy, F, is given by

$$F = E - TS$$

and so

$$S = -F/T + E/T \tag{1.35}$$

Thus, equations 1.34 and 1.35 lead to the expression

$$F = -NkT \ln f \tag{1.36}$$

or alternatively

$$f = e^{-F/RT} \tag{1.37}$$

We shall see later that partition functions are very important in calculations of chemical equilibria, simply because the equilibrium constant can be related to $\exp(-\Delta F/RT)$. They occur again, for a similar reason, in the transition-state theory of chemical kinetics.

f can be calculated if all the energy levels available to the system are known. Supposing that we can calculate these levels theoretically, or measure them spectroscopically, inspection of the preceding 37 equations will show that we can also calculate E, F, S, etc., and show how they vary with T, V and P. In fact, almost any thermodynamic function is calculable in principle from a knowledge of molecular parameters; and so we have effected the unification of quantum mechanics and thermodynamics for which we were looking.

Factorisation of the Partition Function

So far we have said very little about the nature of the energy of molecules except that it is quantised. In general, the total energy of a system is made up of translational, rotational, vibrational and electronic components, and in order to calculate the over-all partition function, we would have to be able to write down every possible combination of these—an impossible task. Fortunately, however, we can factorise the partition function provided that the various energy components are independent of each other, which is a fairly good approximation.

For example, this procedure requires that the vibrational energy levels of a molecule remain the same no matter how fast the molecule is rotating—we shall see later that this is not strictly true, but the error introduced is very small. However, it is strictly true that all forms of internal energy (i.e. rotational, vibrational and electronic) are independent of translational motion.

The total energy of a system is

$$\epsilon_{t,r,vib,e} = \epsilon_t + \epsilon_r + \epsilon_{vib} + \epsilon_e \tag{1.38}$$

where t, r, vib and e denote translation, rotation, vibration and electronic energy, respectively, and each subscript is associated with a quantum number that can take all values 1, 2, 3, ... Thus

$$\begin{aligned} f &= \sum_{t,r,vib,e} g_t g_r g_{vib} g_e \, e^{-\epsilon_{t,r,vib,e}/kT} \\ &= \sum_{t,r,vib,e} g_t g_r g_{vib} g_e \, e^{-(\epsilon_t + \epsilon_r + \epsilon_{vib} + \epsilon_e)/kT} \\ &= \sum_t g_t \, e^{-\epsilon_t/kT} \times \sum_r g_r \, e^{-\epsilon_r/kT} \times \sum_{vib} g_{vib} \, e^{-\epsilon_{vib}/kT} \times \sum_e g_e \, e^{-\epsilon_e/kT} \\ &= f_t f_r f_{vib} f_e \end{aligned} \tag{1.39}$$

Thus the total partition function can be expressed as a product of the partition functions for the individual energy components, and we shall see that these can be calculated fairly readily.

The Translational Partition Function

If we have a particle of mass m confined in a rectangular closed vessel of dimensions a, b and c, the available energy levels are given by

$$\epsilon_{i,j,l} = \frac{h^2}{8m}\left(\frac{i^2}{a^2} + \frac{j^2}{b^2} + \frac{l^2}{c^2}\right) \tag{1.40}$$

where the quantum numbers i, j, l can each take the values 1, 2, 3, . . ., and the degeneracies g_i, g_j and g_l are all unity. Hence,

$$\begin{aligned} f_t &= \sum_{i=1}^{\infty} \sum_{j=1}^{\infty} \sum_{l=1}^{\infty} e^{-\left(\frac{i^2}{a^2} + \frac{j^2}{b^2} + \frac{l^2}{c^2}\right)\frac{h^2}{8mkT}} \\ &= \sum_{i=1}^{\infty} e^{-\left(\frac{i^2}{a^2}\frac{h^2}{8mkT}\right)} \sum_{j=1}^{\infty} e^{-\left(\frac{j^2}{b^2}\frac{h^2}{8mkT}\right)} \sum_{l=1} e^{-\left(\frac{l^2}{c^2}\frac{h^2}{8mkT}\right)} \end{aligned} \tag{1.41}$$

The three terms in equation 1.41 are identical apart from the quantities a, b and c, and we need only evaluate one of them, as follows:

$$\sum_{i=1}^{\infty} e^{-\frac{i^2}{a^2}\frac{h^2}{8mkT}} = \sum_{i=1}^{\infty} e^{-i^2 w^2} \quad \text{where} \quad w = \frac{h}{2a\sqrt{(2mkT)}}$$

If a is of the order of 1 cm and $T = 300$ K, these energy levels are extremely close together ($w^2 \sim 10^{-16}$ erg molecule^{-1}) and we can, without any loss of accuracy, replace the summation by an integral, i.e.

$$\sum_{i=1}^{\infty} \mathrm{e}^{-i^2 w^2} \simeq \int_0^{\infty} \mathrm{e}^{-i^2 w^2}\, \mathrm{d}i$$

(Cf. discussion on rotational partition function on p. 60).

Using the standard integral

$$\int_0^{\infty} \mathrm{e}^{-\alpha^2 x^2}\, \mathrm{d}x = \sqrt{(\pi)}/2\alpha$$

we obtain

$$\sum_{i=1}^{\infty} \mathrm{e}^{-i^2 w^2} = \sqrt{(\pi)}/2w = \frac{\sqrt{\pi}}{2}\frac{2a}{h}\sqrt{(2mkT)} = \frac{a}{h}\sqrt{(2\pi mkT)}$$

Hence, from equation 1.41,

$$f_\mathrm{t} = \frac{a}{h}\sqrt{(2\pi mkT)}\,\frac{b}{h}\sqrt{(2\pi mkT)}\,\frac{c}{h}\sqrt{(2\pi mkT)} = V\left[\frac{2\pi mkT}{h^2}\right]^{\frac{3}{2}} \qquad (1.42)$$

where V is the volume of the rectangular vessel within which the particle of mass m is confined. In fact, it can be shown that this formula is equally valid if the vessel is not rectangular.

The Boltzmann constant k, first introduced as an arbitrary constant in equation 1.4, can be related to easily measurable quantities by means of equation 1.42. From classical thermodynamics we have

$$P = -\left(\frac{\partial F}{\partial V}\right)_{T,N}$$

Using equation 1.36 for F, we obtain

$$P = NkT\left(\frac{\partial \ln f}{\partial V}\right)_T \qquad (1.43)$$

If we consider a monatomic gas such as helium (for which $f_\mathrm{r} = f_\mathrm{vib} = f_\mathrm{e} = 1$, see p. 119), the over-all partition function is just equal to f_t, which, from equation 1.42, is simply a constant multiplied by V. Equation 1.43 therefore becomes

$$P = NkT\left(\frac{\partial \ln V}{\partial V}\right)_T = \frac{NkT}{V}$$

or

$$PV = NkT \tag{1.44}$$

But we know experimentally that $PV = RT$, where R is the gas constant, and so k is simply equal to R/N.

The Specific Heat of Crystals

The theory of the specific heat of crystals, which was given by Einstein in 1907, is of considerable historical importance. Planck had postulated a few years earlier that the emission of radiation must be quantised in order to explain the properties of black-body radiation. Einstein's treatment of the crystal gave us one of the first experimental proofs that energy is quantised, because, as we shall see, the specific heat of a crystal at low temperature can only be explained by means of both quantum and statistical mechanics.

We assume that the total energy of a crystal is made up of two parts, i.e. E_{int}, the internal energy of the N atoms, and a further amount E_{vib} which represents the vibrational energy of the lattice. The internal energy of the atoms is constant, so that the required specific heat at constant volume is

$$c_V = \left(\frac{\partial E_{\text{vib}}}{\partial T}\right)_V \tag{1.45}$$

If each atom executes simple harmonic motion about its mean position, independent of the motions of the other atoms, the vibrational energy of each atom along any one of the three coordinate axes (see p. 65) is

$$\epsilon_{\text{vib}} = (\nu + \tfrac{1}{2})\,h\omega, \qquad \nu = 0, 1, 2, \ldots \tag{1.46}$$

Thus, the total vibrational energy resolved along any axis is

$$E_{\text{vib}} = \sum_{\nu} n_{\nu}\epsilon_{\text{vib}} \tag{1.47}$$

Equations 1.32 and 1.46 lead to the expression

$$n_{\nu} = \frac{N\,\mathrm{e}^{-(\nu+\frac{1}{2})h\omega/kT}}{\sum_{\nu}\mathrm{e}^{-(\nu+\frac{1}{2})h\omega/kT}}$$

and so

$$E_{\text{vib}} = \sum_{\nu} \frac{N\,\mathrm{e}^{-(\nu+\frac{1}{2})h\omega/kT}}{\sum_{\nu} \mathrm{e}^{-(\nu+\frac{1}{2})h\omega/kT}} (\nu + \tfrac{1}{2})h\omega$$

$$= \frac{N \sum_{\nu} (\nu + \frac{1}{2})h\omega\, \mathrm{e}^{-(\nu+\frac{1}{2})h\omega/kT}}{\sum_{\nu} \mathrm{e}^{-(\nu+\frac{1}{2})h\omega/kT}}$$

This result applies only to one dimension; the corresponding expression for three dimensions is

$$E_{\text{vib}} = 3N \frac{\sum_{\nu} (\nu + \frac{1}{2})h\omega\, \mathrm{e}^{-(\nu+\frac{1}{2})h\omega/kT}}{\sum_{\nu} \mathrm{e}^{-(\nu+\frac{1}{2})h\omega/kT}} \tag{1.48}$$

Cancelling $\mathrm{e}^{-\frac{1}{2}h\omega/kT}$ from numerator and denominator gives

$$E_{\text{vib}} = 3N \frac{\sum_{\nu} (\nu + \frac{1}{2})h\omega\, \mathrm{e}^{-\nu h\omega/kT}}{\sum_{\nu} \mathrm{e}^{-\nu h\omega/kT}}$$

$$= 3N \left[\frac{\sum_{\nu} \nu h\omega\, \mathrm{e}^{-\nu h\omega/kT}}{\sum_{\nu} \mathrm{e}^{-\nu h\omega/kT}} + \tfrac{1}{2}h\omega \frac{\sum_{\nu} \mathrm{e}^{-\nu h\omega/kT}}{\sum_{\nu} \mathrm{e}^{-\nu h\omega/kT}} \right]$$

$$= 3N \frac{\sum_{\nu} \nu h\omega\, \mathrm{e}^{-\nu h\omega/kT}}{\sum_{\nu} \mathrm{e}^{-\nu h\omega/kT}} + \tfrac{3}{2}Nh\omega \tag{1.49}$$

The second term in equation 1.49 is the zero-point energy of the entire lattice vibration ($\nu = 0$). The first term is the additional vibrational energy possessed by the lattice at finite temperature T. The two summations in equation 1.49 are easily evaluated in closed form as follows:

$$\sum_{\nu} \mathrm{e}^{-\nu h\omega/kT} = 1 + \mathrm{e}^{-h\omega/kT} + \mathrm{e}^{-2h\omega/kT} + \ldots = \frac{1}{1 - \mathrm{e}^{-h\omega/kT}}$$

$$\left(\text{cf. } 1 + x + x^2 + \ldots = \frac{1}{1-x}\right)$$

Also,

$$\sum_{v} vh\omega\, e^{-vh\omega/kT} = h\omega(0 + e^{-h\omega/kT} + 2\, e^{-2h\omega/kT} + 3\, e^{-3h\omega/kT} + \ldots)$$

$$= h\omega\, e^{-h\omega/kT}(1 + 2\, e^{-h\omega/kT} + 3\, e^{-2h\omega/kT} + \ldots)$$

$$= \frac{h\omega\, e^{-h\omega/kT}}{(1 - e^{-h\omega/kT})^2}$$

$$\left(\text{cf. } 1 + 2x + 3x^2 + \ldots = \frac{1}{(1-x)^2}\right)$$

Hence, equation 1.49 becomes

$$E_{\text{vib}} = \frac{3Nh\omega\, e^{-h\omega/kT}}{(1 - e^{-h\omega/kT})} + \frac{3Nh\omega}{2}$$

$$= \frac{3Nh\omega}{(e^{h\omega/kT} - 1)} + \frac{3Nh\omega}{2} \qquad (1.50)$$

Substituting this expression into equation 1.45, we obtain

$$c_V = \frac{3Nh\omega\left(\dfrac{-d(e^{h\omega/kT})}{dT}\right)}{(e^{h\omega/kT} - 1)^2}$$

$$= \frac{3Nh\omega}{(e^{h\omega/kT} - 1)^2}\, \frac{h\omega}{kT^2}\, e^{h\omega/kT}$$

$$= 3Nk\, \frac{(h\omega)^2}{(kT)^2}\, \frac{e^{h\omega/kT}}{(e^{h\omega/kT} - 1)^2} \qquad (1.51)$$

If, on the other hand, we had regarded the atoms as classical oscillators so that all possible energies were accessible, the summations in equation 1.48 would have been replaced by integrals—with the eventual result* that $c_V = 3R$, independent of temperature. In contrast, it is easy to show either analytically or graphically (use units of $kT/h\omega$ as the horizontal axis) that equation 1.51 falls to zero as $T \to 0$. Note also that at high temperatures, when $h\omega/kT$ is very small, equation 1.50 becomes

$$E_{\text{vib}} = \frac{3Nh\omega}{(1 + h\omega/kT + \ldots - 1)} + \frac{3Nh\omega}{2}$$

* See, for example, G. S. Rushbrooke, *Statistical Mechanics*, Clarendon Press, Oxford, 63-65, 1949.

and so

$$c_V = \left(\frac{\partial E_{\text{vib}}}{\partial T}\right)_V \simeq 3R$$

As mentioned above, this is the classical value and is independent of temperature (cf. the law of Dulong and Petit). The behaviour predicted by equation 1.51 is in excellent agreement with experiment, as shown in *Figure 1.1.* Actually, the prediction of behaviour at low temperatures

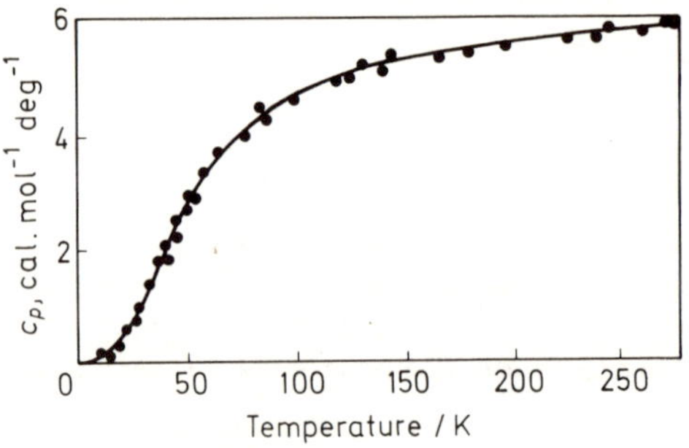

Figure 1.1. Heat capacity of metallic silver as a function of temperature. The heat capacity at constant pressure c_p is the quantity measured, but for solids below room temperature the difference between c_p and c_V is very small

is not perfect, but has been improved to a satisfactory degree by the later work of Debye (1912).

The Correspondence Principle

The equivalence of quantum and classical formulae at high temperatures, which we have just demonstrated, is an example of the correspondence principle. This states that when the separation of the energy levels of a system is very small compared with kT, the system behaves classically. This is beautifully illustrated if we consider the quantal and classical distribution functions for a harmonic oscillator. Classically, a harmonic oscillator spends most time at the end of its traverse, and least time in the middle, where its velocity is greatest. On the other hand, the quantum oscillator in its ground state spends most of its time in the middle (see *Figure 1.2*). However, when we go to more highly excited states, the probability distribution approaches more closely to

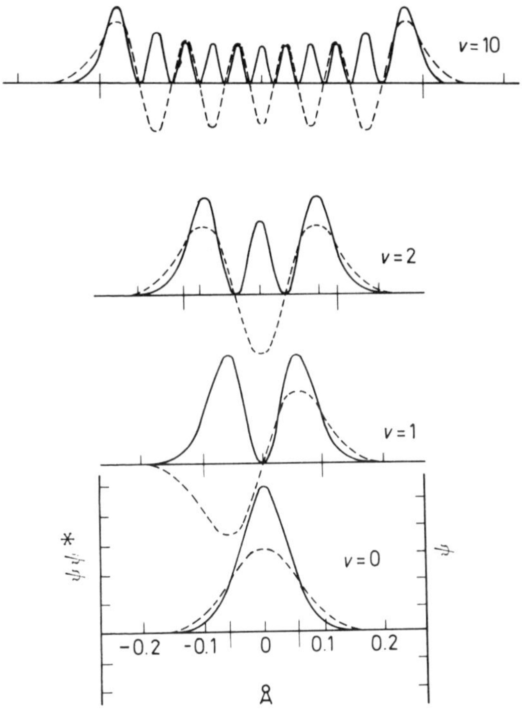

Figure 1.2. Eigenfunctions (broken curves) and probability density distributions (solid curves) of the harmonic oscillator for v = 0, 1, 2 and 10. The abscissae give the displacement for the equilibrium position in Ångstrom units. The eigenfunctions ψ are given by the broken lines and the probability ψψ by the full lines*

the classical one—of course, such 'near-classical' states are only significantly populated at very high temperatures.

The Vibrational Partition Function

The partition function for a one-dimensional oscillator is given by

$$f_{\text{vib}} = \sum_{\nu} g_{\nu}\, e^{-\epsilon_{\nu}/kT} \qquad (1.52)$$

(see Pauling and Wilson, *Introduction to Quantum Mechanics,* McGraw-Hill, 1935), since g_{ν} is unity for all states. Also:

$$\epsilon_{\nu} = h\omega(\nu + \tfrac{1}{2})$$

and so, using the summation on p. 16, we obtain an alternative form of this partition function:

$$f_{\text{vib}} = \frac{e^{-\frac{1}{2}\hbar\omega/kT}}{1 - e^{-\hbar\omega/kT}} \tag{1.53}$$

Either form may be used if care is taken to define the origin from which the energies are measured; but since it is not possible in reality to remove the zero-point energy $\frac{1}{2}\hbar\omega$ from a quantum oscillator, expression 1.53 is more appropriate to experiments made in the laboratory. Very occasionally, vibrational partition functions are calculated with the dissociation limit of the bond as the zero of energy.

Molecules in Fields

We shall see later that spectroscopic measurements are often made on atoms and molecules subjected to intense electric or magnetic fields. Although these fields alter the energies of the atoms or molecules, at equilibrium the population distribution still obeys the Boltzmann law. As a very simple example of the application of Boltzmann's law to molecules in fields, we may consider the problem of the gravitational attraction of the earth for its atmospheric gases.

In addition to their translational, rotational and vibrational energies, the molecules of the atmosphere have a potential energy due to gravitation. Ignoring gravitational sources other than the earth, this potential energy may be expressed as $V(h) = mgh$, where m is the mass of the molecule, h is its height above the earth's surface and g is the acceleration due to gravity. Thus the number of molecules at height h is

$$n(h) = \text{constant} \times e^{-V(h)/kT}$$

Without determining the constant, and using the relation between n and the density of the atmosphere ρ, and its pressure, P, we can write

$$\frac{\rho(h_1)}{\rho(h_2)} = \frac{P(h_1)}{P(h_2)} = \frac{n(h_1)}{n(h_2)} = e^{-[V(h_1) - V(h_2)]/kT}$$

$$= e^{-mg(h_1 - h_2)/kT} = e^{-Mg(h_1 - h_2)RT} \tag{1.54}$$

which is the so-called barometric equation. The real situation is, of course, complicated by temperature and velocity gradients.

Exercises

1. Show either graphically (using units of $kT/h\omega$ as the horizontal coordinate) or analytically (by careful expansion of equation 1.51) that the specific heat of a crystal falls to zero at absolute zero, and tends to a constant value of $3R$ at high temperatures.
2. Show that α, which occurs in equation 1.31, is equal to μ/kT, where μ is the chemical potential.
3. Using either equation 1.29 or equation 1.32, find an expression for the ratio n_j/n_i of two populations as a function of ϵ_i, ϵ_j, g_i, g_j and T. If $g_i = g_j$ and $\epsilon_j > \epsilon_i$, to what temperatures do the following population ratios correspond: $n_j/n_i = 0$; $n_j/n_i = 0.5$; $n_j/n_i = 1$; $n_j/n_i = 1.001$; $n_j/n_i = \infty$? Comment on those ratios which do not yield a meaningful answer for T.
4. Investigate the approximate numerical magnitudes of the partition functions for translation and vibration by means of equations 1.42 and 1.53. After reading Chapter 4, estimate the numerical magnitude of a typical rotational partition function using equation 4.5. Thus, given that $f_e \simeq 1$ (see Chapter 9), estimate the order of magnitude of a typical total partition function by means of equation 1.39.
5. Show that, measured from a zero which corresponds to the dissociated atoms at rest, the vibrational partition function is $(1 - e^{-h\omega/kT})^{-1}\ e^{+\epsilon_d/kT}$ where ϵ_d is the dissociation energy of the molecule.

Chapter Two

ATOMIC SPECTRA—A RÉSUMÉ

Observation of Spectra

The observation and interpretation of atomic and molecular spectra have been major factors in the development of modern quantum mechanics. It was realised in the second half of the nineteenth century that chemical elements and compounds possessed characteristic spectra, and so determination of the positions and relative intensities of spectral lines was undertaken. Quantitative interpretation began when Balmer demonstrated that a simple relation existed between the wavelengths of the lines observed in the visible part of the atomic spectrum of hydrogen. Shortly afterwards Planck introduced the idea that radiant energy was quantised and the very important relation between the frequency, ω, and the energy, E, of a light quantum

$$E = h\omega \tag{2.1}$$

Subsequently, Bohr proposed that a set of stationary energy states is available to an atom, the transition from one level ϵ_i to another level ϵ_j being accompanied by the emission or absorption (or transfer, if a suitable mechanism is available) of a quantum of energy $|\epsilon_i - \epsilon_j|$. Emission occurs if $\epsilon_j > \epsilon_i$, and vice versa for absorption, but in either case the frequency of the electromagnetic radiation is

$$\omega = \frac{|\epsilon_i - \epsilon_j|}{h}$$

These postulates form the basis of the spectroscopic method for obtaining quantitative information about the separation of energy levels in atoms and molecules. We must then turn to quantum mechanics to determine the actual energy levels and which transitions between them are allowed. In this way the pair of levels responsible for each line can be identified. Unfortunately, we must assume that various parameters such as masses, moments of inertia, force constants, etc., are known for the system under investigation. This is not often the case, and so quantum mechanics is used to predict the *relationship* between the wavelengths or frequencies of a series of spectral lines. It is usually possible to work backwards from the frequencies of these lines to calculate the parameters which govern them.

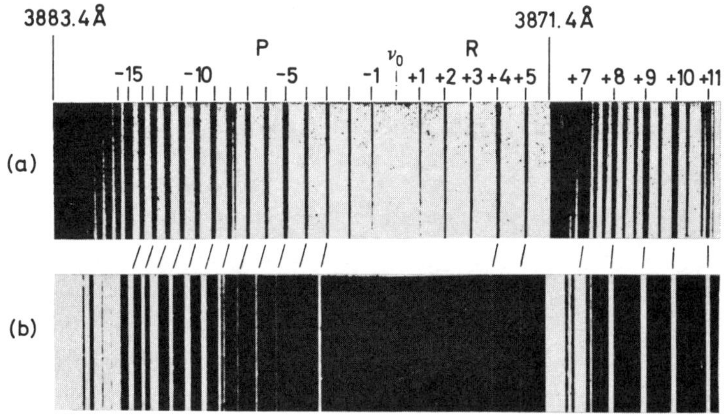

Figure 2.1. Part of photographic (a) emission (b) absorption spectrum of CN *radical in the near ultraviolet*

There is a fundamental difference between the techniques used for observing spectra in emission and in absorption. To record an emission spectrum the sample is incorporated in the source and emission is excited by thermal, electrical or chemical means. The emitted radiation is broken down into its spectral components in a spectrometer and the dispersed frequencies are detected by means of a photographic plate or photoelectric device. Absorption spectroscopy differs in that a continuous source of radiation is used and the sample is interposed between the source and the spectrometer. Photographic records of parts of the absorption and emission spectra of the CN radical are shown in *Figure 2.1*.

The Electromagnetic Spectrum

The region of the electromagnetic spectrum which is of particular relevance to the subject matter of this book extends in wavelength from 1000 Å to 30 m. The units in which wavelength λ is reported vary according to the region considered. Also, radiation is often described in terms of its frequency ω, which is related to wavelength by the equation

$$\lambda\omega = c = 2.998 \times 10^{10} \text{ cm s}^{-1}$$

where c is the velocity of light. Frequency is a particularly useful scale because it is directly proportional to energy (see equation 2.1). Furthermore it is independent of the propagating medium—a point of some significance, since spectrometers are operated under different conditions. Far ultraviolet and far infrared spectrometers are evacuated, whereas normal atmospheric conditions are usually employed in other regions. Velocity and wavelength are affected by the medium but

$$\omega = \frac{c_{\text{vac}}}{\lambda_{\text{vac}}} = \frac{c_{\text{air}}}{\lambda_{\text{air}}}$$

and

$$\lambda_{\text{vac}} = n_{\text{air}} \lambda_{\text{air}}$$

where n_{air} is the refractive index of air. Fortunately, conversion tables are available.

Unfortunately, the expression of natural frequency in cycles s^{-1}, i.e. hertz (Hz), involves large numbers, except at radiofrequencies, where the megahertz is a convenient unit. Consequently, in the ultraviolet and particularly in the infrared regions the wavenumber unit, cm^{-1}, is used:

$$\bar{\upsilon}_{\text{vac}}(\text{cm}^{-1}) = \frac{1}{\lambda_{\text{vac}}(\text{cm})}$$

Hence, the Planck relation may be written

$$E = hc\bar{\upsilon}_{\text{vac}} \tag{2.2}$$

Consider now an atom or molecule emitting radiation of frequency 3×10^{10} Hz, i.e. 1 cm^{-1}. Equation 2.2, together with the value of Planck's constant h, gives

$$E = 6.625 \times 10^{-27} \times 3 \times 10^{10} = 1.99 \times 10^{-16} \text{ erg molecule}^{-1}$$
$$= 2.84 \times 10^{-3} \text{ kcal mol}^{-1}$$

We infer from this observation that the atom or molecule has a pair of energy levels separated by 2.84 cal mol^{-1}. Alternatively, we could have converted the energy-level separation into units of electronvolts (eV); an electron accelerated through a potential of 1 V acquires a kinetic energy equivalent of 1.59×10^{-12} erg molecule^{-1}, and so in the example above

$$E = 1.23 \times 10^{-4} \text{ eV}$$

Table 2.1 summarises the principal units used in various parts of the spectrum and includes also equivalent energies. Additionally, spectroscopists are being asked to adhere to a new unit of wavelength, the nanometre (nm), which is equal to 10^{-9} m or 10 Å.

Table 2.1 TYPICAL WAVELENGTH, FREQUENCY AND ENERGY VALUES

	U.V.	*Visible*	*Infrared*	*Microwave*	*Radiofrequency*
λ(Å)	2 000	5 000	100 000		
λ(μm)			10	10 000	
λ(mm)				10	
λ(m)					10
ω(Hz)	1.5×10^{15}	6×10^{14}	3×10^{13}	3×10^{10}	3×10^{7}
ω(MHz)					30
$\bar{\upsilon}$(cm^{-1})	50 000	20 000	1 000	1	
E(kcal mol^{-1})	142	57	2.8	2.8×10^{-3}	2.8×10^{-6}
E(eV)	6.2	2.5	0.12	1.2×10^{-4}	1.2×10^{-7}

A spectrum is characterised not only by the positions of its spectral lines but also by their shapes and intensities. Most line contours approximate to Lorentz functions when recorded on a linear frequency scale. Thus the contour height H of a line at a point $\pm\Delta\omega$ from its centre is given by

$$H = \frac{a}{(\Delta\omega)^2 + b^2}$$

The peak height at the line centre ($\Delta\omega = 0$) is therefore given by

$$H_{\text{max}} = a/b^2$$

and the two points on the contour at half peak height, $H_{\text{max}}/2$, are separated in frequency by an amount $\Delta\omega_{\frac{1}{2}} = 2b$. Ideally, therefore, H_{max} and $\Delta\omega_{\frac{1}{2}}$ should be reported along with the position of a line. $\Delta\omega_{\frac{1}{2}}$ is independent of the number of atoms or molecules involved in the emission or absorption process, whereas peak height is clearly not independent. It is therefore desirable (though not always possible) to report data for standard peak heights.

For many purposes the peak height is a sufficient measure of intensity, but strictly the intensity is proportional to the area under the contour of the line, i.e. the integrated intensity. There is, however, an important difference in the relationship between intensity and concentration of the sample for emission and absorption spectra. In emission from optically thin sources, intensity is directly proportional to the number of atoms or molecules emitting radiation; provided this number can be determined, molar or molecular standard intensities can readily be obtained. In absorption spectroscopy the relation between concentration and intensity of light transmitted by the sample, I, follows Beer's law:

$$\frac{I}{I_0} = \exp(-n\epsilon')$$

where I_0 is the intensity of incident light, n is the number of absorbing atoms or molecules per unit area of incident beam and ϵ' is a constant for a given line called the extinction coefficient. A plot of the optical density, $D = \log_{10}(I_0/I)$, against frequency enables the gram-atomic or gram-molecular extinction coefficients to be determined. This procedure can be very tedious since many commercial spectrometers trace out the percentage of light, a, transmitted by the sample, i.e.

$$\frac{I_0}{I} = \frac{100}{a}$$

and consequently the results usually have to be replotted in terms of optical density. Furthermore, in all experiments additional complications can arise from the distortion of line contours due to instrumental imperfections and the use of spectrometer slits of finite width.

The discussion above applies both to molecular and atomic spectroscopy but we shall now devote the remainder of this chapter to atomic spectra.

Selection Rules

The atomic spectra about which we know most are those of hydrogen and the hydrogen-like atoms, i.e. all those atomic systems which contain only one electron. All the available energy levels in such an atom can be calculated by solving the non-relativistic Schrödinger equation, and this enables us to write down the energies, and therefore the wavelengths, of all possible transitions (see *Figure 2.2*). Then, as

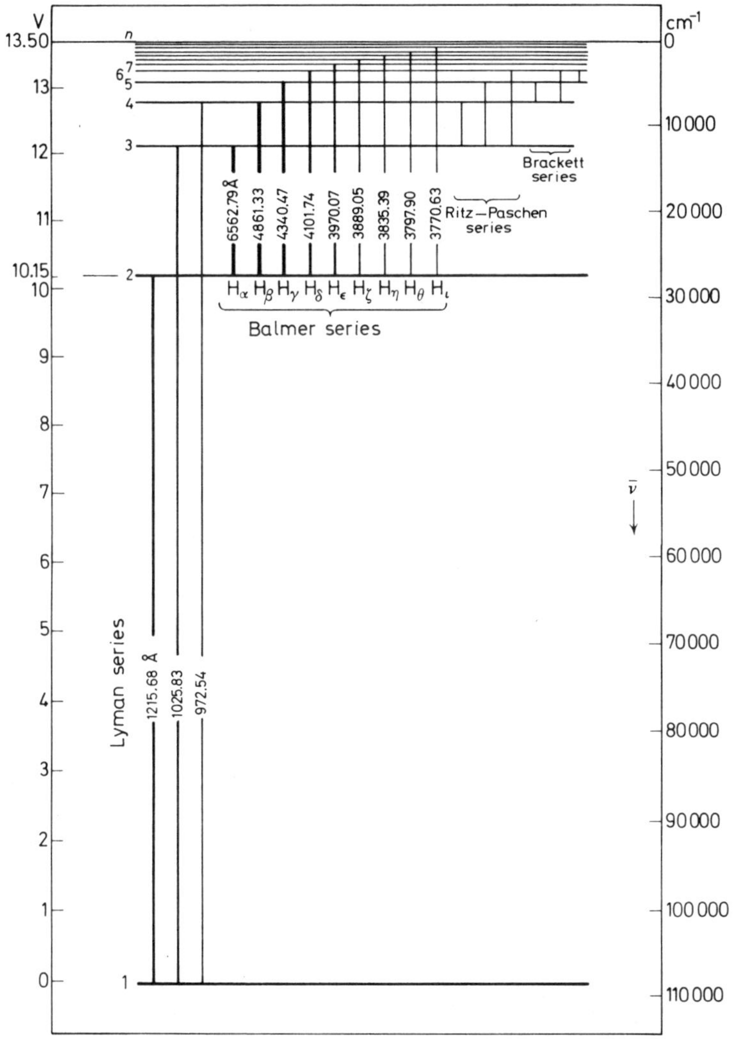

Figure 2.2. The energy level scheme of hydrogen

shown for hydrogen in *Figure 2.3,* we can construct series of predicted spectra. These spectra give only the positions of the lines and a complete comparison with experiment requires calculation of the intensity of each line. *Figure 2.4* shows an experimental record of one of the most intense series in the H-atom spectrum, and this should be

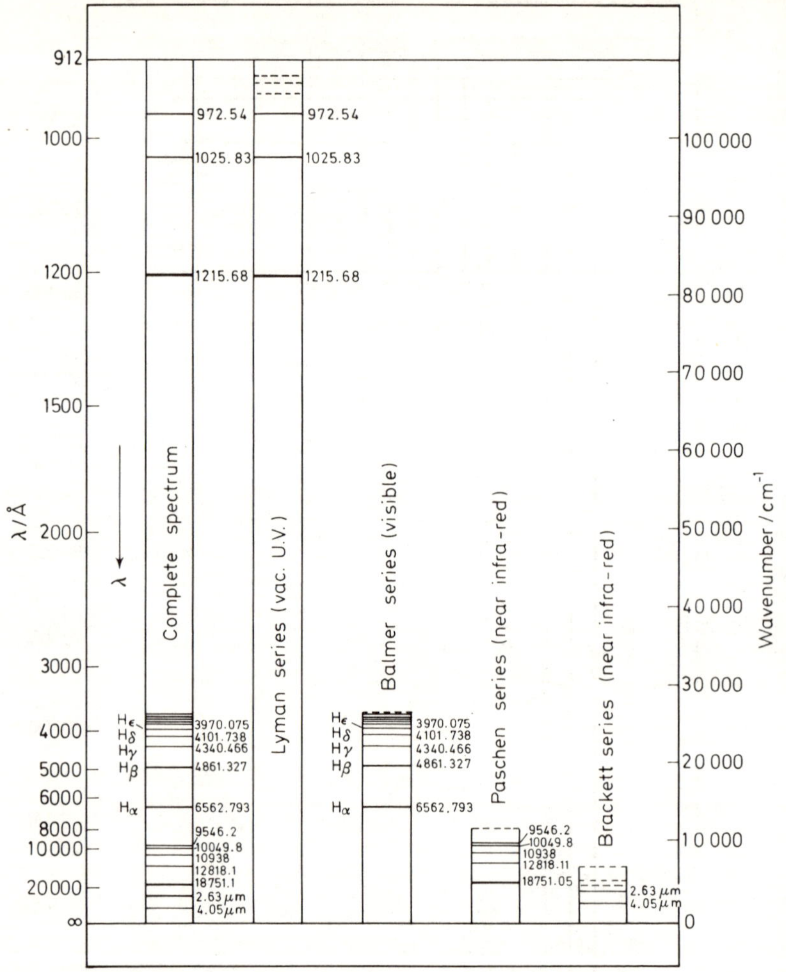

Figure 2.3. The spectral lines of the hydrogen spectrum

compared with the relevant part of *Figure 2.3.* However, the final analysis of a spectrum is not quite as simple as this. Under very high resolution, each member of the Balmer series shown in *Figure 2.4* is seen to be not a single line, but a *multiplet.* The reason for this is that in constructing *Figures 2.2* and *2.3* we solved only the non-relativistic Schrödinger equation. When due allowance is made for the fact that the effective mass of the electron is a function of its energy (velocity), we

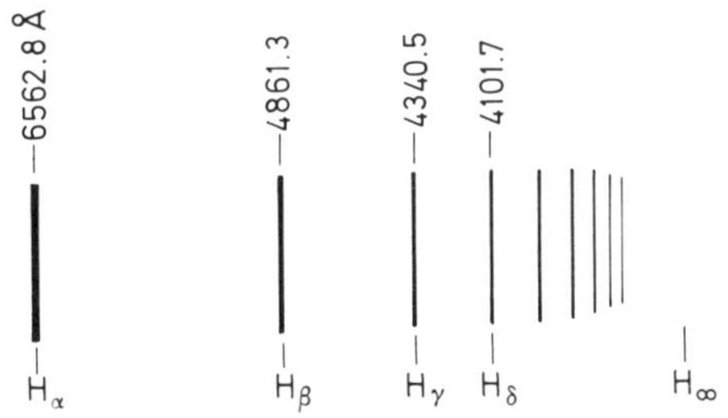

Figure 2.4. Emission spectrum of the hydrogen atom in the visible and near ultraviolet region (Balmer series). H_∞ gives the theoretical position of the series limit

find that levels with a different subsidiary quantum number, l (s, p, d, . . .), are no longer precisely degenerate for a given value of the principal quantum number, n. This fact enables the fine structure of the spectrum to be analysed, and the energy level transitions which are found to occur are shown in *Figure 2.5.* We find that transitions can take place between any value of n*, but the quantum number l must obey the selection rule

$$\Delta l = \pm 1$$

All other transitions are *forbidden.* It is interesting to note that Δn may equal zero and we then observe the transition between a 2p and a 2s energy level, which is the 3-cm extragalactic radiation.

The reasons for selection rules are well understood in quantum mechanics, but the details are beyond the scope of this book. A treatment of the interaction of an electromagnetic wave with an atom, using time-dependent perturbation theory, leads to the following results:

(1) An atom in state ϵ_i will only absorb energy from the wave and be raised to a new state ϵ_j if the frequency of the radiation is such that $\epsilon_j - \epsilon_i = h\omega$.

(2) The probability that an atom, initially in state ϵ_i, will find itself in state ϵ_j after a time t is proportional to t and to the quantity $(\int \psi_i \mu \psi_j \, d\tau)^2$, where μ is the 'instantaneous dipole moment' of the atom, and ψ_i, ψ_j are the wave functions of the states. (A hydrogen

* n, the principal quantum number, principally determines the size and energy of the orbital; l, the azimuthal quantum number, determines the shape, and $l \leqslant (n - 1)$; m is the magnetic quantum number. Lower case symbols are used when referring to single electron states and capitals are used for whole atoms containing more than one electron.

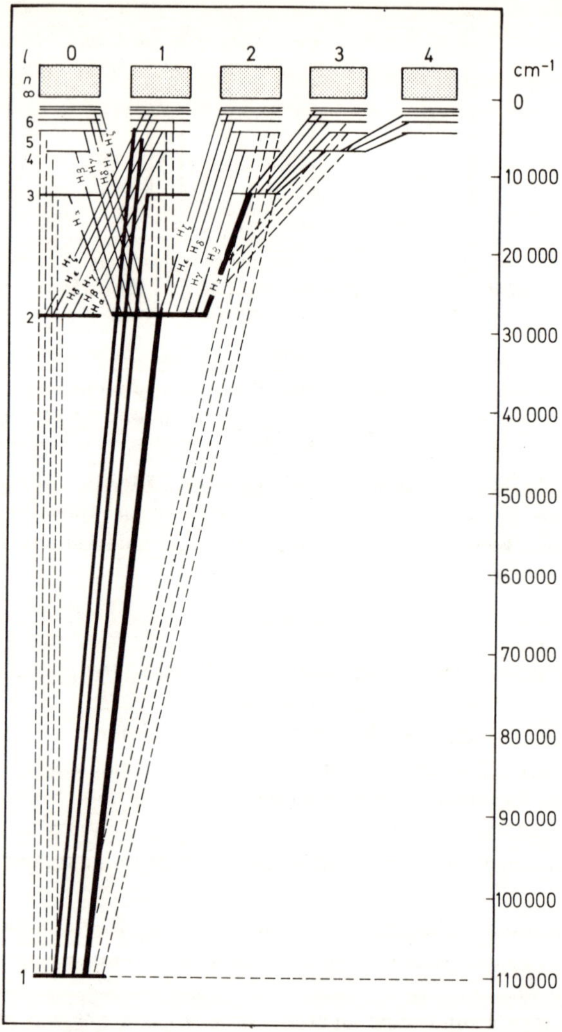

Figure 2.5. Energy level diagram of the H *atom, including fine structure. The Balmer series is indicated with* H_α, H_β, H_γ, *etc. The broken lines refer to forbidden transitions*

atom has no observable dipole moment because the quantity $\int \psi_i \mu \psi_i \, d\tau$ is zero for any state of the system, despite the fact that μ, the instantaneous dipole, is always the dipole arising from a proton and an electron separated by a distance r, i.e. $\mu = er$.) The integral $\int \psi_i \mu \psi_j \, d\tau$ is called the *dipole matrix element* of the transition. The rules for emission are complementary to those rules for absorption. It is an easy matter to evaluate these integrals, e.g. $\int\psi_{1s}\mu\psi_{2s}\mathrm{d}\tau$, $\int\psi_{1s}\mu\psi_{2p}\mathrm{d}\tau$, etc., for a hydrogen atom (see any textbook of quantum mechanics), and it turns out that most of them are zero; they are only non-zero for transitions between states which obey the selection rules

$$\Delta l = \pm 1; \qquad \Delta m = 0, \pm 1$$

As (2) above indicates, these are the only transitions with a finite probability.

Our spectra have demonstrated the first of these selection rules; the second rule can only be verified by studying the effect of a magnetic field on a spectrum, i.e. the Zeeman effect. Under these conditions states of different magnetic quantum number m have slightly different energies and the fine structure of the spectrum is therefore elaborated. A detailed analysis shows that the Δm selection rule is obeyed.

For atoms other than hydrogen or hydrogen-like atoms, no exact calculations are available to help analyse the observed spectra. Many thousands of atomic spectral lines have been reported* and the problem of assigning each line to a transition between two states is analogous to solving a very complicated jig-saw puzzle. However, a high degree of order has been established by trial-and-error methods, and we will now discuss one or two of the simplest cases; the others are similar but more complicated.

After the one-electron atoms, the spectrum of helium is the simplest to study. The observed lines (*Figure 2.6*) appear to fall into two distinct sets of hydrogen-like series which are quite independent of each other—so much so that early investigators thought there were two different kinds of helium, which they called ortho- and para-helium. However, the true explanation is that the transitions on the left-hand side of the diagram are all between states in which the two electrons are paired—called *singlet* states—and those on the right-hand side are between states in which the electrons are unpaired—*triplet* states. No transitions are observed between states of different multiplicity,† and so we may define a third selection rule:

$$\Delta S = 0$$

* See, for example, Moore, *Atomic Energy Levels,* volumes I, II and III.

† If $2S$, where S is the spin quantum number, is the number of unpaired spins in the atom, the *multiplicity* of the state is given by $(2S + 1)$, i.e. 1 when $2S = 0$ and 3 when $2S = 2$).

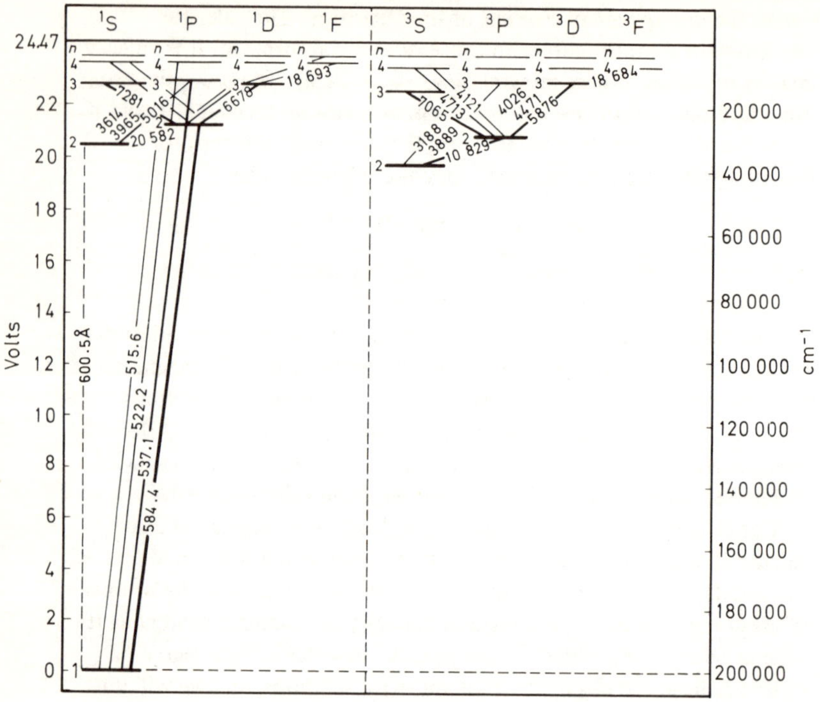

Figure 2.6. Energy level diagram for helium. The series in the visible and near ultraviolet regions correspond to the transitions between terms with n ⩾ 2

As a consequence of this the 2^3S state is a *metastable* state, i.e. it is the lowest of the triplet states, and cannot revert to the ground 1^1S state by radiation (this is forbidden by the ΔS selection rule) but can only do so by collision processes. (The lettering S, P, D, F, . . ., is used to denote the different total orbital angular momentum states of the atom, analogous to s, p, d, f, . . . for the H atoms; thus 2S denotes a 1s2s atom, 3P denotes 1s3p, etc.—the occurrence of states like $2s^2$ is so rare that no confusion can arise.) The ions Li^+, Be^{2+}, etc., have similar spectra.

Next in order of complexity comes the alkali metal lithium. The spectra of all the alkali metals are very similar to each other, and *Figure 2.7* shows the energy level diagram for potassium. Each of these atoms has one electron outside a closed shell, and to a first approximation might be expected to resemble a hydrogen-like system. However, one important fact stands out—that the very small energy difference

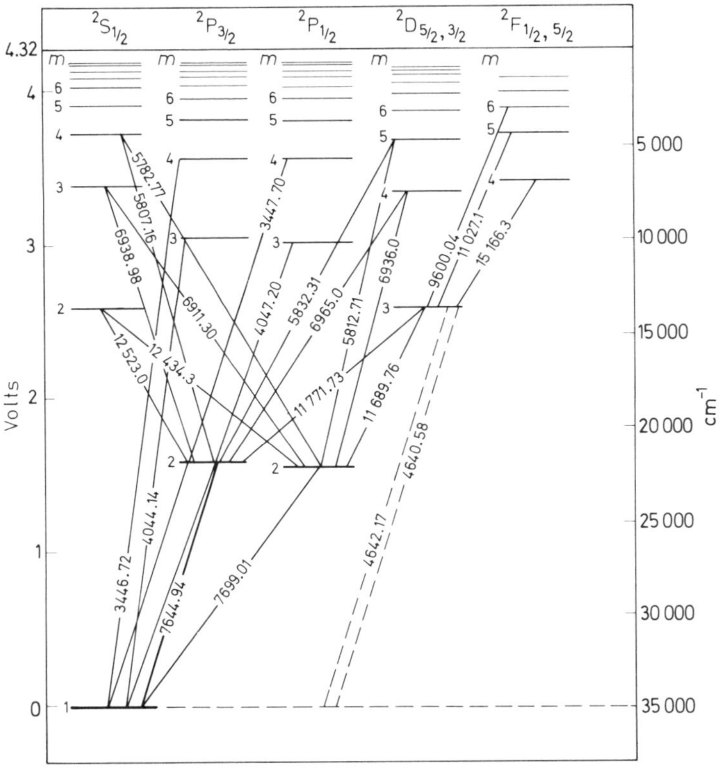

Figure 2.7. Energy level diagram for potassium

between the 2s and 2p levels of the hydrogen atom has now become very large indeed (i.e. compare the levels 2^2S and 2^2P on the diagram, which would correspond to $1s^2 2s$ and $1s^2 2p$ for lithium). Also, it is apparent that the ^{2}P levels are all split (also the ^{2}D and ^{2}F levels, but the splittings are smaller). The interpretation of this splitting is that it represents different alignments between the resultant orbital angular momentum ($L = 1$ in this case) and the resultant spin angular momentum ($S = \frac{1}{2}$). In other words, a knowledge of the quantum numbers L and S is not sufficient to define the state uniquely, and an examination of all the spectral data available shows that we need to define another quantum number J, which can take the values $J = |L + S|, |L + S - 1|, |L + S - 2|, \ldots, |L - S|$, to describe all possible such alignments. For the alkali metals in their ground states, therefore, we have just the two values of J, $\frac{3}{2}$ and $\frac{1}{2}$, so that the ^{2}P state is split

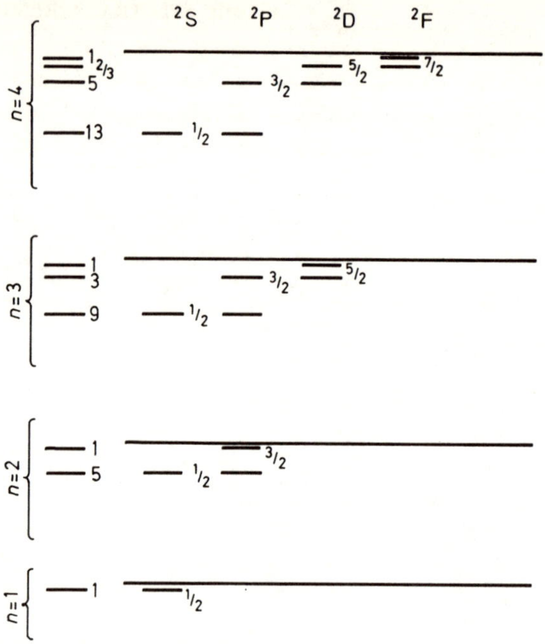

Figure 2.8. Fine structure of hydrogen-like energy levels. The value of J is given on the right-hand side of the levels

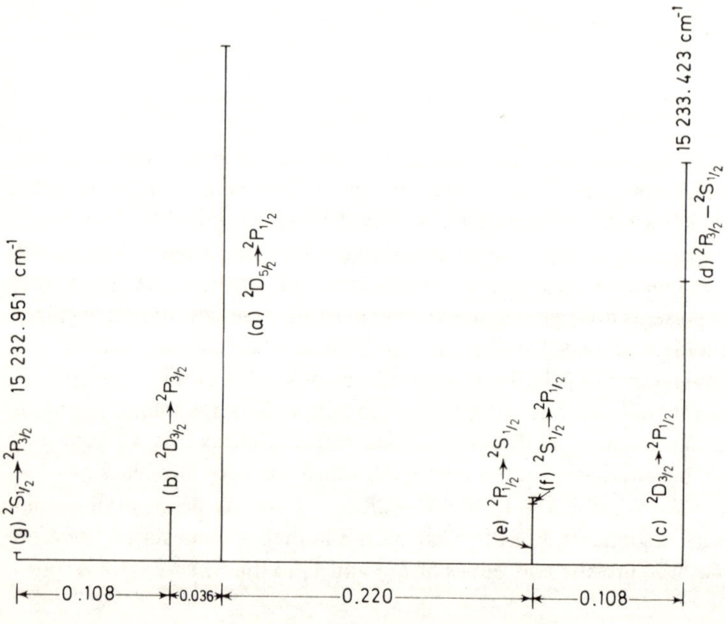

Figure 2.9. Theoretical fine structure of H_α *line*

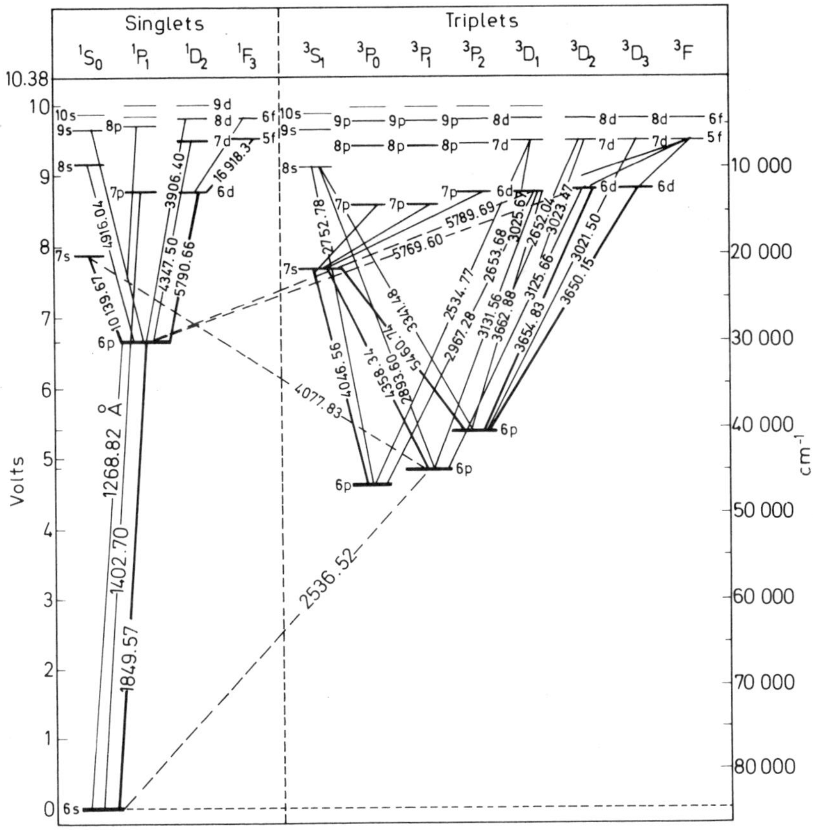

Figure 2.10. Energy level diagram of Hg. *The wavelengths of the more intense* Hg *lines are given*

into $^2P_{\frac{3}{2}}$ and $^2P_{\frac{1}{2}}$. This gives rise to the well-known doublet structure of the alkali metal spectra. Furthermore, it is J that appears to be important in determining the selection rules, since we find that the only transitions to occur are those for which

$$\Delta J = 0, \pm 1 \text{ (but not } J = 0 \rightarrow J = 0)$$

This is the most general form of the selection rule for atomic spectra.

We ought now to go back and look at H and He to see how the J selection rule works. In discussing hydrogen, we made no mention of the fact that a ^2P state (e.g. 2p) could be either $^2P_{\frac{3}{2}}$ or $^2P_{\frac{1}{2}}$, although we did mention the splitting between 2s and 2p. The J splitting amounts to about 0.3 cm^{-1} for the 2p level and 0.1 cm^{-1} for the 3p level. The reason that our previous description appeared to be consistent with experimental observation is that levels with the same values of n and J have the same energy (*Figure 2.8*), a fact that comes out of the full relativistic treatment of the hydrogen atom if the spin is treated in a proper manner. *Figure 2.9* then shows the fine structure of one of the Balmer lines, and demonstrates that even what appear to be single lines in the multiplet have a composite structure.

Turning now to helium, we see that the J selection rule does not prevent transitions between singlet and triplet states, and, in general, these are not *strictly forbidden.* However, they are usually weak, so much so in the helium case that they have not been observed. A well-known example of a singlet-triplet transition is the 2537 Å line of Hg–this is due to the transition $6^3P_1 \rightarrow 6^1S_0$ (see *Figure 2.10*). Although much weaker than the line at 1850 Å, $6^1P_1 \rightarrow 6^1S_0$, it is nevertheless sufficiently strong to be the cornerstone of modern photochemistry.

Chapter Three

ROTATIONAL SPECTRA OF MOLECULES AND THE TECHNIQUES OF SPECTROSCOPY

Rotational Spectra of Diatomic Molecules

The spectra of molecules are considerably more complicated than those of atoms. In atoms we can have only electronic transitions, but in molecules there are transitions due to changes in rotational, vibrational or electronic energy, and, in general, we have a superposition of all three. It would be logical to discuss first the electronic spectra of molecules, but we would not get very far before we needed to know the properties of vibrational and rotational transitions. The simplest approach therefore is to discuss the spectra in increasing order of their transition energies. A further difficulty that arises in the molecular case is that the rotational and vibrational spectra are, in general, very complex, except for diatomic molecules and a few small, symmetrical polyatomic molecules.

To begin with, we will regard our diatomic molecule as a *rigid* dumb-bell of length r separating the two nuclear masses m_1 and m_2 (*Figure 3.1*). This is a good approximation, since the nuclear mass is so much greater than the electronic mass. For example, in H_2 99.9% of the total mass is concentrated in the two nuclear regions, each about

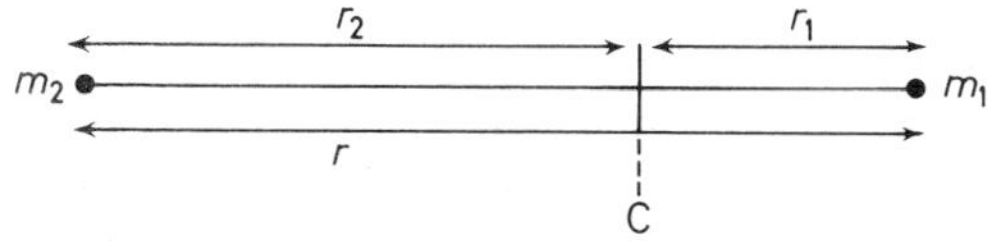

Figure 3.1. Rigid dumb-bell model of a diatomic molecule

10^{-13} cm in diameter, and these two nuclei are separated by a distance of the order of 10^{-8} cm. The centre of mass of the molecule is at C, and the system will rotate about an axis which passes through C. In classical mechanics, the energy of rotation is

$$E_r = \tfrac{1}{2}I\omega_a^2 (\equiv P^2/2I \text{ where } P = I\omega_a) \tag{3.1}$$

where I is the moment of inertia and ω_a is the angular velocity (i.e. $\omega_a = 2\pi\omega$, where ω is the frequency of rotation in hertz).

Now

$$I = m_1 r_1^2 + m_2 r_2^2$$

where

$$r_1 = \frac{m_2 r}{m_1 + m_2} \quad \text{and} \quad r_2 = \frac{m_1 r}{m_1 + m_2}$$

so

$$I = \frac{m_1 m_2^2}{(m_1 + m_2)^2} r^2 + \frac{m_1^2 m_2}{(m_1 + m_2)^2} r^2$$

$$= \frac{m_1 m_2}{m_1 + m_2} r^2 = \mu r^2 \tag{3.2}$$

where μ is called the reduced mass of the system. Thus the molecule is equivalent to a mass μ rotating on a circle of radius r about a fixed axis at C, perpendicular to the internuclear axis.

In the special case of a diatomic molecule, there can be no meaning to rotation about the internuclear axis. All the mass is concentrated on this axis, so that r (and therefore I) is zero for this particular rotation.

It is fairly easy to solve the Schrödinger equation for an effective mass μ rotating on a circle of radius r (cf. the treatment of the 'rigid rotator' in any textbook of quantum mechanics). The allowed energy levels prove to be

$$\epsilon_J = \frac{h^2}{8\pi^2 \mu r^2} J(J+1) = \frac{h^2}{8\pi^2 I} J(J+1) \tag{3.3}$$

where J is a quantum number which can take all positive integral values including zero. Thus we have a series of levels with a progressively increasing spacing between them (*Figure 3.2*). There are actually $(2J + 1)$ degenerate functions for each value of J, a fact which will be important when we return to the discussion of the statistical mechanics of rotating molecules.

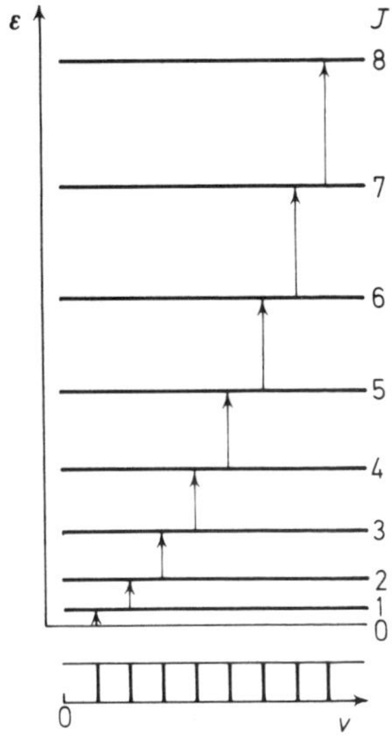

Figure 3.2. The energy level diagram and the resulting far infrared spectrum of a rigid rotor

To find the selection rules for rotational transitions, we again need to evaluate the integrals $(\int \psi_i \mu \psi_j d\tau)^2$. We see immediately that if $\mu = 0$ (i.e. the dumb-bell has no dipole moment) the intensity is zero, and this rule is rigorously obeyed. For molecules which do have a dipole moment, the evaluation of these integrals shows that only transitions corresponding to $\Delta J = \pm 1$ are allowed. Therefore, in absorption, the spectrum consists of a series of equally spaced lines (*Figure 3.2*). If we define the *rotational constant*

$$B = h/8\pi^2 Ic \tag{3.4}$$

the spacing of these lines may be deduced from the selection rule

$$J' - J'' = +1$$

where we use the convention that $'$ represents the upper state and $''$ represents the lower state.

The frequency $\bar{\upsilon}$ (in wave-number units) of any line is given by

$$hc\bar{\upsilon} = \Delta\epsilon$$

i.e.

$$\bar{\upsilon} = \frac{\epsilon' - \epsilon''}{hc} \tag{3.5}$$

so that

$$\bar{\upsilon} = B[(J'' + 1)(J'' + 2) - J''(J'' + 1)]$$

$$\bar{\upsilon} = 2B(J'' + 1) \qquad (3.6)$$

where J'' can take the values 0, 1, 2, . . ., since this is the lower state in the transition. Thus, our series of equidistant lines has a spacing of $2B$ cm^{-1}. As an example, we may consider the molecule HCl^{35}, which is known to have a bond length of 1.27 Å from independent electron diffraction measurements. Substitution in equation 3.4 then gives

$$B = \frac{6.6 \times 10^{-27}}{8\pi^2 \times 3 \times 10^{10} \times \frac{35}{36} \times 1.66 \times 10^{-24} \times (1.27 \times 10^{-8})^2} = 10 \text{ cm}^{-1}$$

Therefore, the spectrum will consist of a series of lines at 20, 40, 60, 80, 100, . . . cm^{-1}, or 0.5, 0.25, 0.166, 0.125, 0.1, . . . mm wavelength. This is in the far infrared and difficult to observe, falling between the limits of the easily accessible infrared (say 400 cm^{-1}) and the easily accessible microwave (say 1 mm) regions. Nevertheless, the rotational spectrum of HCl has been obtained, the lines of which do show this pattern (*Table 3.1, Figure 3.3*).

Table 3.1 ABSORPTION SPECTRUM OF HCl IN THE FAR INFRARED ($J' = J'' + 1$)

$J'' \rightarrow J'$	$\bar{\upsilon}_{obs}$(cm^{-1})	$\Delta\bar{\upsilon}_{obs}$(cm^{-1})	$\bar{\upsilon}_{calc} = 20.89J' - 1.6 \times 10^{-3}J'^3$(cm^{-1})
$0 \rightarrow 1$	20.88	20.88	20.89
$1 \rightarrow 2$	41.75	20.87	41.67
$2 \rightarrow 3$	62.57	20.82	62.43
$3 \rightarrow 4$	83.39	20.82	83.16
$4 \rightarrow 5$	104.13	20.74	103.85
$5 \rightarrow 6$	124.30	20.17	124.49
$6 \rightarrow 7$	145.03	20.73	145.08
$7 \rightarrow 8$	165.51	20.48	165.60
$8 \rightarrow 9$	185.86	20.35	186.04
$9 \rightarrow 10$	206.38	20.52	206.40
$10 \rightarrow 11$	226.50	20.12	226.65

It is immediately obvious that *provided the molecule has a dipole moment,* the measurement of the rotational spacings is a very powerful tool for determining bond lengths, and, because of this fact, the bond lengths of most heteronuclear diatomic molecules are more precisely known than those of homonuclear diatomics such as H_2, F_2, etc. (*Table 3.2*). We have seen that for fairly light molecules such as HCl, the *pure rotational spectrum* occurs in the (relatively inaccessible)

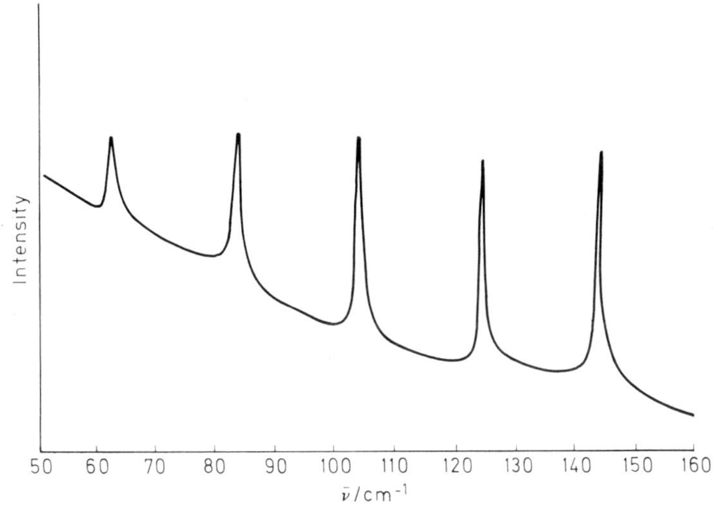

Figure 3.3. Far infrared spectrum of HCl *vapour obtained from an interferogram*

Table 3.2 MOLECULAR PARAMETERS FROM MICROWAVE SPECTROSCOPY

Molecule	B_e(MHz)	r_e(Å)	μ(D)*
HCl^{35}	317 510	1.275	1.18
FCl^{35}	15 483.69	1.628	0.88
$Br^{79}Cl^{35}$	4 570.92	2.138	0.57
ICl^{35}	3 422.30	2.321	0.65
$C^{12}O^{16}$	57 897.5	1.128	0.11
$N^{14}O^{16}$	51 084.5	1.151	0.16
$C^{12}S^{32}$	24 584.3	1.535	2.0

* D = debye.

far infrared. However, when the molecules have rather larger moments of inertia, the spectrum occurs in the range accessible by microwave techniques. Thus, two quite different experimental techniques are needed to span the whole range of rotational spectra.

Infrared Spectrometers

Commercial infrared spectrophotometers are sensitive, reliable instruments which record spectra (usually on precalibrated chart paper) under semiautomatic conditions. The operator simply selects the region to be scanned and inserts the sample into the appropriate cell or holder, and the recording of the spectrum is then automatically controlled. Wavelengths from 2 to 200 μm, i.e. from 5000 to 50 cm^{-1}, can be covered by a single instrument, but it is more usual to have a second machine for the far infrared region, generally considered to be >40 μm, i.e. <250 cm^{-1}. Inevitably, the instrumentation required to produce such a spectrometer in a compact and versatile form is very complex, but the basic principles of operation are simple. We shall ignore the mechanical and electrical problems and concentrate on the optical features.

Single- and Double-beam Spectrometers

Emission spectrometers and single-beam absorption spectrometers have a simple and rather similar layout, as shown in *Figure 3.4.* In absorption, elaborate precautions must be taken to remove the absorption

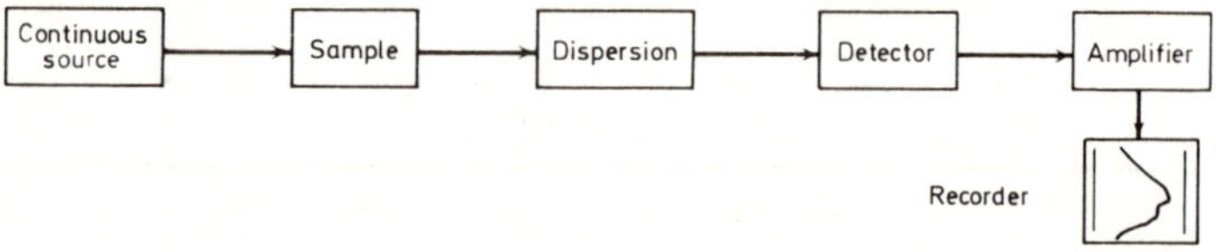

Figure 3.4. Block diagram of a single-beam spectrometer

spectrum of air and compensate for the spectral profile of the source, otherwise a 'background' spectrum must be subtracted from the observed trace in order to obtain the spectrum of the sample. The introduction of double-beam instruments eliminated this problem and proved tremendously useful to the chemist, who often needs to obtain the spectrum of a sample in solution. A typical layout for a double-beam absorption spectrometer is shown in *Figure 3.5.*

Radiation from the source is split into two beams, one passing through the sample and the other serving as a reference (in the case of a solution the reference beam would be passed through the pure solvent). An oscillating mirror assembly, or rotating sector mirror, reflects the two beams alternately into the monochromator and they are detected by a thermocouple. The resulting alternating signal is examined, and any difference between the sample and reference signals (e.g. due to the absorption of the solute in the case of a solution) is used to actuate a servomechanism which drives an attenuator into the reference beam.

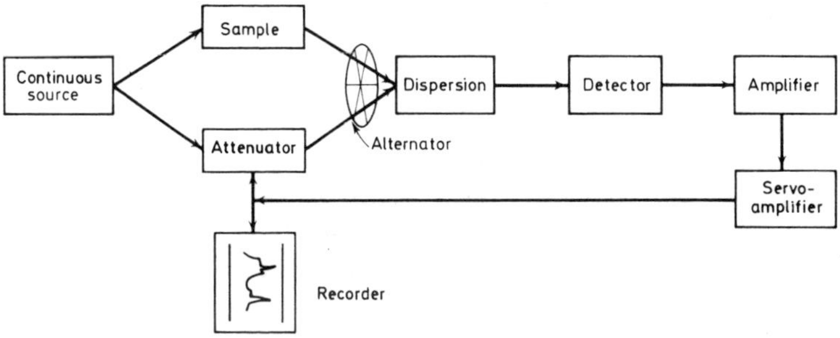

Figure 3.5. Block diagram of a double-beam absorption spectrometer

The mechanism is arranged to maintain an optical null-point balance between the two beams entering the monochromator (i.e. to keep the amplitude of the alternating signal zero). The position of the attenuator is converted mechanically into 'percent transmission' of the sample (e.g. the solute in a solution) and recorded as a function of wavelength or wavenumber on a pen recorder. Apart from cancelling the absorption due to the solvent, the double-beam technique automatically compensates for the absorption of air and the spectrum of the source.

To obtain optimum versatility, an instrument usually consists of a basic unit comprising slit mechanism, cell housing, mirrors, filters and pen recorder, etc., which will accept a range of sources, detectors, and prisms or gratings to complete the spectrometer. Components are selected according to the spectral range under investigation.

Sources

The most common sources for the excitation of emission spectra include open arcs or sparks, conventional gaseous discharge tubes, or electrodeless cavities which are excited by radiofrequency generators. In absorption spectroscopy it is essential to have a source which provides a continuous band of wavelengths. The emission from a hot body at 1000–1600 K reaches a peak at $\sim 2\ \mu m$ and extends well into the infrared with diminishing intensity. Thus, commercial sources for the range 2–50 μm are usually electrically heated rods composed of sintered thorium and cerium oxides (Nernst Glower) or of silicon carbide (Globar). High-pressure mercury arcs enclosed in quartz envelopes are superior in the far infrared; at wavelengths greater than 100 μm the output is approximately double that of a Globar.

Dispersing Optics

The polychromatic radiation emitted from, or passed through, the sample under investigation must be dispersed into its constituent wavelengths. This is achieved by either refraction or diffraction of the light beam. Separation of the spectral lines by refraction requires a prism and there is a practical limitation to this method set by the fact that no suitable prism materials are available for use at wavelengths greater than 50 μm. Gratings do not suffer from this disadvantage and consequently they are particularly useful in the far infrared. Improvements in ruling techniques have resulted in the normal use of gratings throughout the infrared region. However, we are still faced with the problem of using several gratings or prisms to cover the whole range. Suitable prism materials are listed in *Table 3.3,* together with the spectral ranges over which they are generally used. Windows which are

Table 3.3

	Spectral range	
Lithium fluoride	2.5–7 μm	4 000–1 400 cm^{-1}
Sodium chloride	6–14 μm	1 700– 700 cm^{-1}
Potassium bromide	12–25 μm	800– 400 cm^{-1}
Caesium iodide	25–40 μm	400– 250 cm^{-1}

transparent to infrared radiation (and sometimes lenses) are also required in a spectrometer. Sodium chloride, potassium bromide and caesium bromide or iodide are widely used for these purposes at wavelengths up to 50 μm, but in the far infrared region high-density polyethylene is the most satisfactory material.

In a spectrograph the spectrum is recorded on a photographic plate mounted in the focal plane. There are no moving parts involved and all the spectral lines register simultaneously. This is, therefore, a particularly accurate method for determining the positions of lines. Unfortunately, the method is tedious and is unsatisfactory for measuring intensities and line contours. In most commercial instruments a monochromator is coupled to a thermoelectric or photoelectric detector. The essential feature of a monochromator is that an exit slit is mounted in the focal plane to isolate a narrow region (the width of the slit being determined by the resolution required) of the spectrum. The radiation passes through the slit on to a detector. A spectrum can be scanned by moving the slit and detector unit across the focal plane of the instrument, but more usually the prism or grating is rotated so that the spectral lines are swept across the stationary slit.

Detectors

Photocells with lead sulphide, lead selenide or indium antimonide targets have a limited application in the near infrared at wavelengths less than 7 μm. They are notable for speed of response and sensitivity. Thermocouples and bolometers are widely used over the entire infrared region but some difficulties are encountered in the far infrared, due, in part, to the reduced absorbing power of the black coating on the sensitive surface. A more uniform response is obtained from the Golay pneumatic detector, which can be used throughout the infrared region. The target of this detector is a small gas-filled chamber which contains an evaporated layer of aluminium to absorb radiation. When radiant energy falls on it the pressure in the chamber rises and distorts the back wall, which acts as an external mirror. This mirror focuses an auxiliary beam of light on to a photocell, and the amount of light falling on the cell is determined by the distortion of the mirror. Thus, the output of the photocell is indirectly coupled to the infrared radiation falling on the Golay cell.

Interferometers

One of the more recent advances in instrumentation has been the application of interferometers to obtain very weak emission and absorption spectra. They have proved particularly useful in the far infrared region, where conventional spectrometers are handicapped by the low energy output of available sources and the lack of sensitivity of detectors. The most important advantages of an interferometer compared with a normal spectrometer are (1) that all wavelengths present in the radiation fall on the detector for the total observation time; this improves the signal-to-noise ratio of the output from the detector, and (2) the axial symmetry of an interferometer allows circular entrance and exit holes to be used (instead of narrow slits) to collimate the light passing through the instrument. The second factor greatly enhances the amount of light collected by the detector.

A typical Michelson interferometer for absorption spectrometry is shown in *Figure 3.6.* Instead of the monochromatic source used in the determination of the velocity of light, a high-pressure mercury arc S is used as a source of continuous wavelengths in the far infrared region.

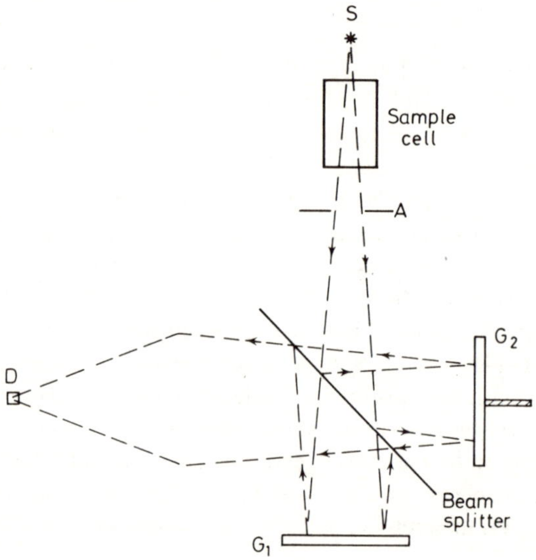

Figure 3.6. Layout of a Michelson interferometer: S, mercury arc; A, circular aperture; D, Golay detector; G_1, fixed optical flat; G_2, movable optical flat

The radiation is collimated, passed through the sample cell and the circular entrance aperture A, and divided by a thin sheet of polyethylene into two beams which fall on optically flat mirrors G_1 and G_2. After reflection at G_1 and G_2 the beams come together again and are directed on to the detector D. The two beams interfere constructively or destructively depending on the frequency of the component radiation and the path difference X between the two beams. The mirror G_2 is mounted on a precision screw so that X can be varied by means of a mechanical drive.

If a monochromatic source is used and the cell is evacuated, a simple cosine interferogram is recorded as a function of X. With a continuous source, a complex interferogram is obtained. By carrying out a Fourier transform, using a computer, this interferogram can be converted into the spectrum of the source. With a sample in the cell, the interferogram (and, hence, the absorption spectrum) of the sample superimposed on that of the source can be obtained. Part of the pure rotational infrared spectrum of HCl vapour recorded in this way is shown in *Figure 3.3.*

Microwave Spectrometers

A simple microwave spectrometer is shown in *Figure 3.7.* The absorption cell is an evacuable length of wave-guide which is separated from the rest of the wave-guide circuit by two vacuum-tight 'windows', usually made of thin mica sheet. The microwave power is generated by

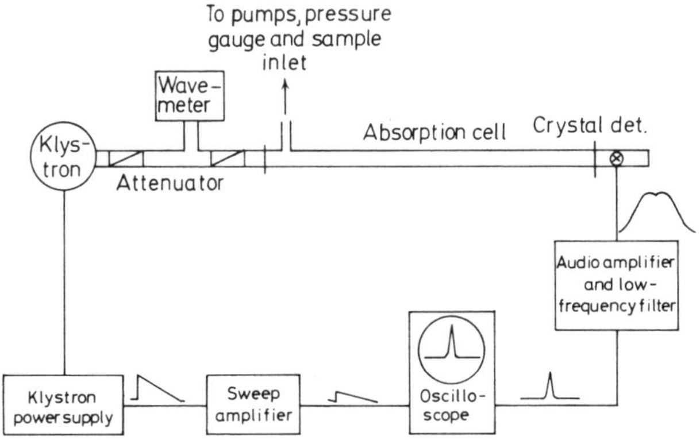

Figure 3.7. Block diagram of a microwave spectrometer

an electron tube known as a klystron, and the transmitted radiation is detected by a suitable crystal and amplifier. The simplest klystron is tunable over a small frequency range by varying the anode voltage, and if the anode is driven by an oscilloscope time-base the horizontal sweep of the oscilloscope spot is then equivalent to a frequency scale. Thus, it is an easy matter to display on the screen a plot of transmitted power against frequency. Then, by tuning the cavity wavemeter into resonance at the same frequency, it is a simple matter to determine the wavelength to about 1 part in 10^4. The block diagram of a much more sophisticated instrument is shown in *Figure 3.8.* In this particular instrument, the frequency sweep is geared to the chart drive of an automatic pen recorder. More important, however, is that frequencies are determined by electronic beat methods. WWV is the U.S. Bureau of

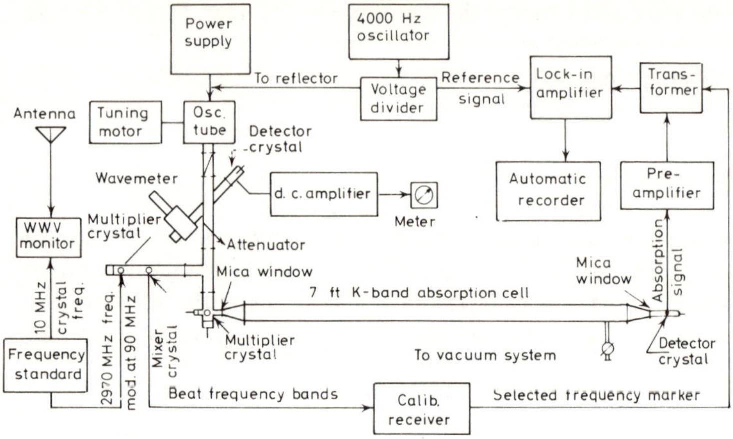

Figure 3.8. Block diagram of a millimetre-wave spectrometer employing a phase-sensitive amplifier with source modulation and automatic recording

Standards 10 kHz transmission, whose frequency is accurate to 1 part in 10^9. In Britain the Droitwich carrier wave of 200 kHz is used as the frequency standard. By generating harmonics, this fundamental frequency is multiplied up until a frequency in the experimental range is obtained, and, of course, this is still accurate to 1 part in 10^9. The absorption frequencies are then compared with this standard by beat methods, with an accuracy of 1 Hz. The implications of this phenomenal precision will be noted later.

Centrifugal Distortion

Returning to the measured frequencies of the HCl rotational spectrum listed in *Table 3.1*, we see that the lines are not quite equally spaced. This deviation is well outside the experimental error in measuring the wavelengths, and is systematic in the sense that it can be well represented by a simple formula, as shown in the final column. The reason for this is simply centrifugal distortion; i.e., as the molecule rotates, the bond is stretched very slightly, and this, of course, alters the moment of inertia. The molecule is in fact a non-rigid rotor.

If the molecule is a harmonic oscillator, which is a reasonable approximation for molecules in their lowest vibrational state (*Figure 3.9*), the restoring force for an internuclear spacing of r is

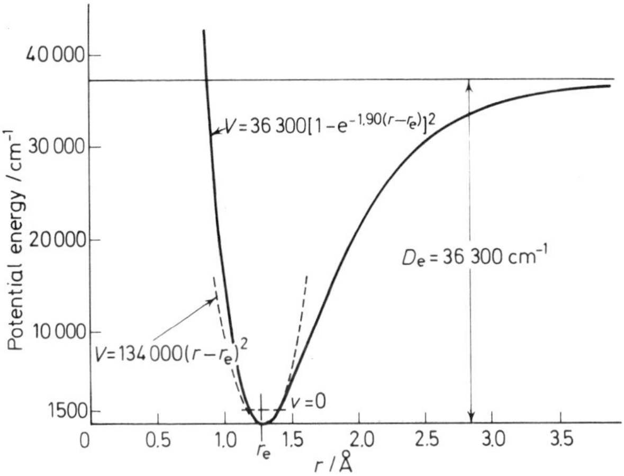

Figure 3.9. The harmonic oscillator (dashed line) and Morse (solid line) potential energy curves for HCl

simply $k(r - r_e)$, where k is known as the force constant; also, for a harmonic oscillator, $k = 4\pi^2 \bar{\upsilon}^2 c^2$, where $\bar{\upsilon}$ is the vibrational frequency in cm^{-1}. Without going into details, the solution of the Schrödinger equation for this system is

$$\epsilon = \frac{h^2}{8\pi^2 \mu r^2} J(J+1) - \frac{h^4}{32\pi^4 \mu^2 r^6 k} J^2(J+1)^2 + \text{smaller terms}$$

$$= hcB\,J(J+1) - hcD\,J^2(J+1)^2 + \text{etc.} \qquad (3.7)$$

where D, the centrifugal distortion constant, is $4B^3/\bar{\upsilon}^2$. For HCl, where $B \sim 10\ cm^{-1}$ and $\bar{\upsilon} \sim 3000\ cm^{-1}$, this works out as $D \sim 4 \times 10^{-4}\ cm^{-1}$, so that the correction in going from equation 3.3 to equation 3.7 is quite small. The selection rules remain unchanged, and it follows that the allowed rotational frequencies are now

$$\bar{\upsilon} = 2B(J'' + 1) - 4D(J'' + 1)^3 = 2BJ' - 4DJ'^3 \qquad (3.8)$$

compared with the experimentally found expression

$$\bar{\upsilon} = 20.89J' - 0.0016J'^3 \qquad (3.9)$$

which is given in *Table 3.1.* Thus a simple method of dealing with such experimental data is to plot $\bar{\upsilon}/J'$ against J'^2. Inspection of equation 3.8 shows that this will be a straight line of slope $-4D$ and intercept $2B$, from which B and D may be obtained.

Anticipating the results of the next chapter, we will accept that each vibrational state of a molecule has its own characteristic values of B and D, which we label B_v and D_v, where v is the quantum number of the particular vibrational state. Equation 3.9 represents the frequencies observed for HCl gas at room temperature, i.e. with virtually all the molecules in the ground ($v = 0$) state; hence, we have $B_0 = 10.445 \text{ cm}^{-1}$ and $D_0 = 4.0 \times 10^{-4} \text{ cm}^{-1}$. Going back to equations 3.2 and 3.4, this gives a value of r_0, the bond length of HCl in its ground vibrational state, of 1.276×10^{-8} cm.

Applications of Microwave Spectroscopy

The growth of microwave spectroscopy out of radar research opened up an enormous field of chemical spectroscopy, which is attractive from several points of view. Adequate sources of power are available, particularly near 1 cm, and the wavelength is easily varied and can be measured with great accuracy. The transmitted signals are readily amplified by standard electronic techniques, which means that intensity measurements are relatively easy to make, in contrast with some other fields of spectroscopy, where absolute line intensities are difficult to determine. This combination of advantages leads to a great number of applications, of which we shall mention five.

(a) *Mass abundances.* It is necessary to measure the relative intensity of two lines corresponding to the same transition of different isotopic species of the same molecule to obtain the abundance ratio. This ratio may be obtained more conveniently with a mass-spectrometer.

(b) *Mass ratios.* If we determine the rotational constant B_e (see next chapter for the significance of B_e) for two isotopic species of a diatomic molecule, then

$$B_e^{(1)} = \frac{h}{8\pi^2 c\mu_1 r_e^2}; \qquad B_e^{(2)} = \frac{h}{8\pi^2 c\mu_2 r_e^2}$$

therefore

$$\frac{B_e^{(1)}}{B_e^{(2)}} = \frac{\mu_2}{\mu_1} = \frac{M_2(M+M_1)}{M_1(M+M_2)} \tag{3.10}$$

An accurate measurement of the isotope effect can be used to obtain a precise value for the ratio of the two isotope masses from μ_2/μ_1. The procedure is complicated and beyond the scope of this book, but the interested reader is referred to the work of C. H. Townes *et al.* (*Phys. Rev.*, **73**, 1334, 1948) on ICl.

The method can be generalised to more complicated molecules, and the results for the chlorine atom obtained by a variety of methods are given in *Table 3.4.* These mass ratios are independent of any uncertainties in the values of c, the velocity of light, h, Planck's constant, or r_e, the equilibrium internuclear separation of the molecule.

Table 3.4 MASS RATIOS BY DIFFERENT TECHNIQUES FOR Cl^{35}/Cl^{37}

Cl^{35}/Cl^{37}	*Mass ratio**
Mass spectroscopy	0.945 977 7 ± 20
Nuclear reaction	0.945 989 3 ± 110
Microwave (ICl)	0.945 980 1 ± 50
Microwave (FCl)	0.945 977 5 ± 40
Microwave–molecular beam (KCl)	0.945 980 3 ± 15
Microwave (CsCl)	0.945 978 1 ± 30
Microwave (ClCN)	0.945 990 6 ± 120

* The error limits represent the uncertainties in the least significant digits.

Rotational Constants in Different Vibrational States

We have already mentioned that each vibrational state of a molecule has a slightly different rotational constant B_v and, therefore, slightly different bond lengths. Because of the enormous resolution available, it is simple to find the spectra for various values of v, the vibrational quantum number, provided that a sufficient population of the state in question exists at the experimental temperature. (Again, see the discussion in the next chapter.)

Accurate Bond Lengths

Since the rotational constants of molecules can be measured to a precision approaching nine significant figures, the main errors in the computed bond lengths (using equations 3.2 and 3.4) are due to uncertainties in the assumed values of the fundamental constants h and c, and in the nuclear masses.

Dipole Moments

Because accurate measurements of the line intensity may be made, it is possible to take the theoretical quantum-mechanical expression for the absolute line intensity and work back to the square of the dipole moment, which comes in through the factor $(\int \psi_i \mu \psi_j d\tau)^2$. This is relatively tedious, but it can be used in simple cases to give dipole moments about as good as those obtainable from dielectric constant measurements.

However, much more precise dipole moments can be obtained very easily by using what is known as the *Stark effect*. The absorbing gas is subjected to an electric field (E) of the order of a few thousand volts per centimetre. A typical experimental arrangement is shown in *Figure 3.10*. If the rotating molecule has a dipole moment, its rotation

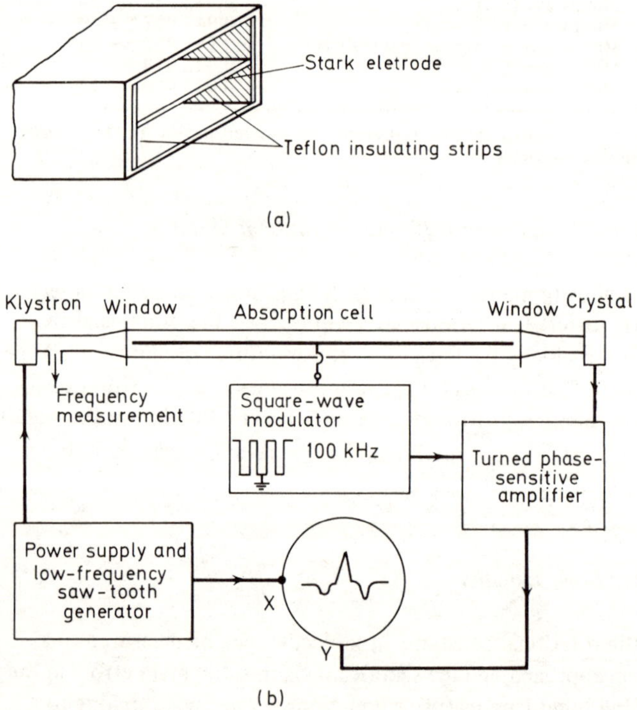

Figure 3.10. Stark modulation spectrometer: (a) the Stark electrode in the absorption cell; (b) a block diagram of the microwave spectrometer

will no longer be uniform—it will accelerate as it comes into alignment with the field and decelerate as it moves out of alignment. This leads to a net change in the energy of rotation.

Quantum-mechanically, we need to evaluate the integral $\Delta E_{J'} = \int \psi_{J'} \mu E \cos\theta \, \psi_{J'} \, d\tau$ and the analogous one for $E_{J''}$ in order to determine the frequency of the transition in the presence of the applied field. The result is that the energy of each level for a linear molecule is changed by an amount which is proportional to $(\mu E)^2$, and is also a function of J, the rotational quantum number, and M^2, where M is the quantum number determining which of the $2J + 1$ possible orientations the molecule in question takes up. Thus, depending upon $|M|$, some levels are moved upward by the field, and others downward. *Figure 3.11* shows some experimental data on the $J = 1$ state of the OCS molecule, plotted so as to demonstrate the proportionality of the splittings to E^2. There is a selection rule that $\Delta M = 0$, so that $\frac{1}{3}$ of the molecules, those with $M = 0$, suffer a decrease in energy with increasing field, whereas $\frac{2}{3}$ of the molecules, i.e. those with both $M = +1$ and

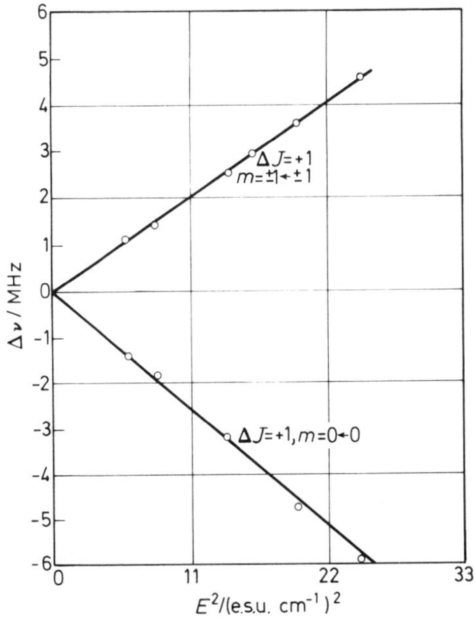

Figure 3.11. Comparison between theory and experimental measurement of Stark effects in the $J = 1 \rightarrow 2$ transition of the linear molecule OCS

$M = -1$, show the opposite behaviour. *Figure 3.12* shows this effect very nicely, for it is clear that, as the applied field increases, the Stark splitting increases until the transition is completely resolved into two components, one twice as intense as the other. The numerical value of

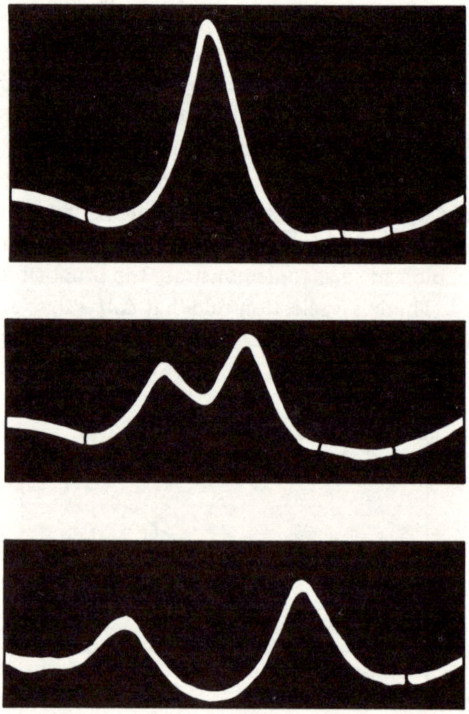

Figure 3.12. Stark splitting of the $J = 1 \rightarrow 2$ rotational line of OCS. *Upper curve, no field; middle curve, 750* V/cm; *lower curve, 1070* V/cm

the dipole moment is obtained from the slope of a plot of the splitting against E^2, usually to quite high precision. Of course, following the discussion above (p. 51), such measurements can be made for any individual vibrational state that can be observed—for example, the dipole moments are known for as many as five different vibrational states of some of the alkali halide molecules. A selection of results obtained by the Stark effect is given in *Table 3.5.* The theory of the Stark effect for non-linear molecules is more complicated, giving ΔE_J proportional to E for degenerate levels (first-order Stark effect) and to

Table 3.5 DIPOLE MOMENTS FROM MICROWAVE SPECTROSCOPY

Molecule	*Dipole moment in ground state (D)*
OCS	0.71
OCSe	0.75
HCN	3.00
ClCN	2.80
N_2O	0.17
HCCCl	0.44
HCCCN	3.6

E^2 for non-degenerate levels (second-order Stark effect). Nevertheless, accurate values of dipole moments are still obtainable.

Rotational Spectra of Non-linear Molecules

For all molecules except the lightest of diatomics, rotational spectra occur in the microwave region simply because of the mass factor, whose effect is shown in *Figure 3.13.* The treatment we have given above is valid for all linear molecules, e.g. OCS, ICN, etc. *Figure 3.14(a)* shows the spectrum of a symmetric-top molecule. A symmetric top is one in which there is a threefold or higher axis of symmetry, e.g. CH_3I, CH_3–Hg–I, CF_3–C≡CH. Two of the three principal moments of

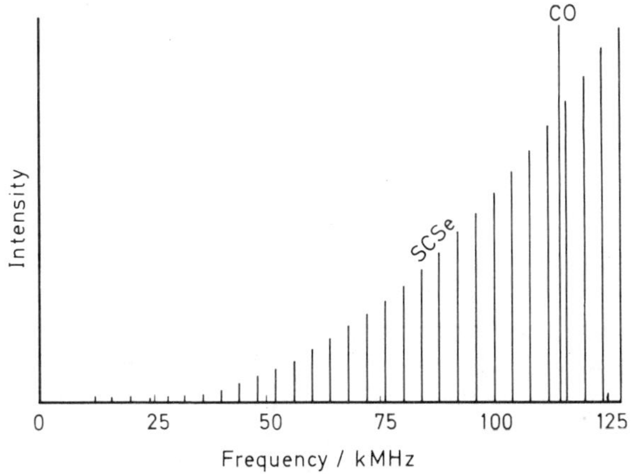

Figure 3.13. Microwave rotational spectrum of a linear molecule with a relatively small (CO) *and a relatively large* (SCSe) *moment of inertia. Only the first rotational line of* CO *is shown*

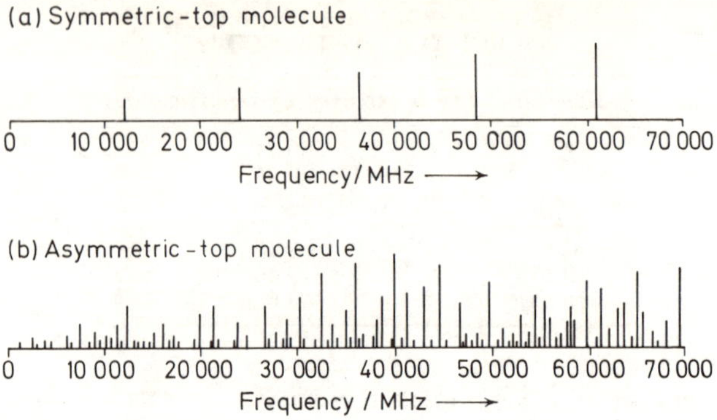

Figure 3.14. Illustration of the general pattern of the microwave rotational spectra of symmetric-top and asymmetric-top molecules

inertia are thus identical. The third principal moment of inertia may be smaller (prolate top) or larger (oblate top) than the other two. The spectrum consists of a set of approximately equally spaced lines, characteristic of the degenerate moment of inertia (i.e. analogous to the spectrum of a linear molecule). Molecules which have three non-equivalent principal moments of inertia are called asymmetric tops. Their spectra are very complicated, and difficult to unravel except by working back from an assumed structure of the molecule, using estimated values of the molecular parameters. *Figure 3.14*(*b*) shows an example of the rotational spectrum of an asymmetric top molecule. Apart from structural data, the most important information derived from the spectra of asymmetric tops relates to hindered rotation about single bonds. Barriers to internal rotation have been determined for a large number of molecules, e.g. CH_3NH_2, CH_3OH, etc.

Hyperfine Structure

An extremely important aspect of microwave spectroscopy is that one can detect the slight alteration of the rotational energy levels due to the magnetic moments of the nuclei. This effect manifests itself as satellites (see, for example, *Figure 3.15*) which are due to interaction between the rotation of the molecule and reorientation of the nuclear spin. Study of these spectra leads to information on the nuclear spins, and about the asymmetry of the electronic environment in the region of

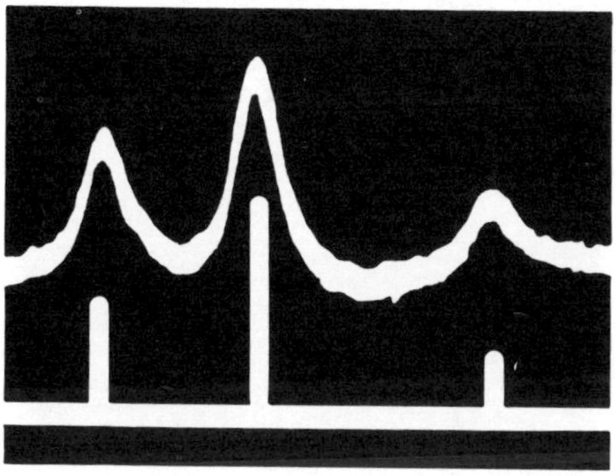

Figure 3.15. Theoretical and observed hyperfine structure of the $J = 0 \rightarrow 1$ rotational transition of DCN

these nuclei—the latter effect has an interesting bearing on theories of valence.

Exercises

1. What are the reduced masses of H_2, D_2, HD, LiH, $H^{35}Cl$, $H^{37}Cl$, $D^{35}Cl$, $D^{37}Cl$? Express your answers in atomic mass units, and convert the answer for D_2 into grams.
2. The rotational constant B for HI is 6.55 cm^{-1}; calculate the moment of inertia of the HI molecule in gram cm^2, its bond length in Å, and the position of the $J = 0 \rightarrow 1$ transition in cm^{-1}, in mm and in MHz.
3. Given that HCN is a linear molecule and that you are able to measure its rotational spectrum: for how many isotopic species of the molecule must you measure B in order to determine both bond lengths? From among only the nuclear masses H = 1, 2; C = 12, 13; N = 14, 15, which combinations will give you the most accurate bond lengths? Finally, if you did not know that the carbon atom was in the middle, how many more isotopic modifications would you need to study to prove that it was?
4. Assuming that CH_3I is a tetrahedral molecule, with C–H = 1.10 Å and C–I = 2.15 Å, calculate its two principal moments of inertia.
5. Using equation 3.9 calculate r_0 and an approximate value for the force constant k for HCl. Compare the answer for k with the more precise result given in Chapter 5.

6. Lines in the pure rotational spectrum of $C^{12}O^{16}$ are observed at

$$7.682, \qquad 11.508, \qquad 15.316, \qquad 19.100\ \text{cm}^{-1}$$

Calculate r_0 and D_0 for the molecule.

Chapter Four

STATISTICAL MECHANICS OF ROTATING MOLECULES

The Boltzmann Distribution

We have seen (equations 3.3 and 3.4) that the allowed energy levels for a rigid rotator are

$$\epsilon_J = Bhc\, J(J+1) \tag{4.1}$$

and that this accounts fully for the observed spectra, apart from a small correction due to centrifugal distortion. The effect of this correction on the statistical mechanics of the rotator is negligible. We have also seen that a rotational energy level E_J has a degeneracy which can be lifted by the Stark effect, and that the number of wave functions, i.e. *a priori* probabilities corresponding to each quantum state, is $2J + 1$. Thus the rotational partition function is

$$f_r = \frac{1}{\sigma}\sum_{J=0}^{\infty} g_J\, e^{-EJ/kT} = \frac{1}{\sigma}\sum_{J=0}^{\infty} (2J+1)\, e^{-BhcJ(J+1)/kT} \tag{4.2}$$

In this equation, σ is the symmetry number, which, for diatomic molecules, takes the value 2 if the two nuclei are identical, but 1 if they are different. It is introduced to take account of the fact that if a homonuclear diatomic molecule is rotated through 180°, the configuration is indistinguishable from the original one, but for a heteronuclear molecule the configurations are different. Also, we may express the population of any rotational state J

$$n_J \;\alpha\; \frac{1}{\sigma}(2J+1)e^{-BhcJ(J+1)/kT} \tag{4.3}$$

Figure 4.1 shows the relative populations, based on $n_{J=0} = 1$, for a

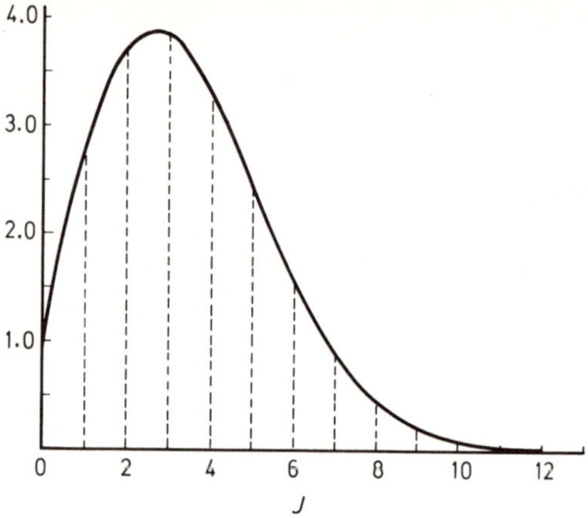

Figure 4.1. Thermal distribution of the rotational levels for T = 300 K *and B = 10.44* cm^{-1} *(that is, for* HCl *in its ground state). The curve represents the function* $(2J + 1)e^{-BhcJ(J+1)/kT}$ *as a function of J. The broken line ordinates give the relative populations of the corresponding rotation levels*

molecule such as HCl. This probability distribution goes through a maximum for HCl at 300 K at about $J = 3$, and it can easily be shown by differentiating equation 4.3 with respect to J that

$$J_{\text{max}} = \left(\frac{kT}{2Bhc}\right)^{1/2} - \tfrac{1}{2}$$

i.e. the value of J at which the population is a maximum increases with T, and increases as the molecule gets heavier (B decreases).

In order to work out the absolute populations, we need to be able to evaluate the summation in equation 4.2. Fortunately, for most molecules, the sum can be approximated by an integral to a high degree of precision at room temperature and above. It is only for light molecules such as H_2 and hydrides at low temperatures that the error is significant. If we write $y = Bhc/kT$, we can put

$$\sum_{J=0}^{\infty} (2J + 1)\,e^{-J(J+1)y} \simeq \int_0^{\infty} (2J + 1)\,e^{-J(J+1)y}\,dJ$$

$$= \int_0^{\infty} e^{-J(J+1)y}\,d[J(J + 1)] = \frac{1}{y}. \qquad (4.4)$$

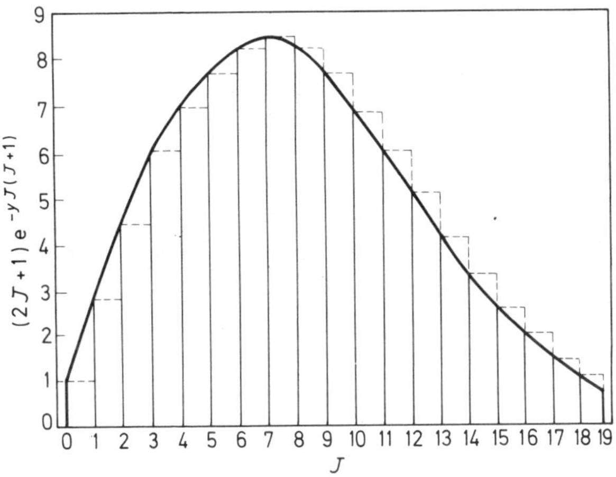

Figure 4.2. Comparison of the sum

$$\sum_{J=0}^{\infty} (2J+1)e^{-yJ(J+1)}$$

and the integral

$$\int_0^{\infty} (2J+1)e^{-J(J+1)y}\,dJ$$

The heights of the vertical lines represent the individual terms $(2J+1)e^{-yJ(J+1)}$. *The area under the smooth curve is the integral; the area under the dashed line is the sum*

The nature of the approximation can be seen by referring to *Figure 4.2.* Thus,

$$f_r = \frac{1}{\sigma y} = \frac{kT}{\sigma hcB} = \frac{8\pi^2 IkT}{\sigma h^2} = \frac{T}{1.4388\sigma B} \tag{4.5}$$

if T is in K and B in cm^{-1} (see equation 3.4).

The exact expression, required at low temperatures, is

$$f_r = \frac{1}{\sigma y}\left(1 + \frac{y}{3} + \frac{y^2}{15} + \ldots\right) \tag{4.6}$$

and it is seen that the approximate expression (equation 4.5) is simply the first term in this expansion. Consequently, using equation 4.5,

$$\frac{n_J}{N} = y(2J+1)\,e^{-J(J+1)y} \tag{4.7}$$

One might add that classical mechanics leads also to equation 4.5 for the partition function, and this is not surprising, since, by using the integral approximation, we have made the assumption that the energy levels are very close together, which is true for high T and small B.

In the last chapter we saw in *Figure 3.13* that there was a variation in intensity in the rotational lines of SCSe, and that this intensity distribution looks remarkably like a part of *Figure 4.2.* In the next chapter we shall see this rotational population distribution showing up even more convincingly when we consider the vibrational spectra of diatomic molecules.

Exercises

1. The bond length of CO is 1.13 Å. Estimate which is the most intense line in its pure rotational spectrum at (i) 100 K; (ii) 1000 K.
2. Use equation 3.9 together with equation 4.3 to calculate the relative intensities at 300 K of the first four rotational lines of HCl.
3. Use the data provided in *Figure 3.13* to estimate the value of J'' for the most intense line in the spectrum of SCSe at 300 K.
4. The bond length of H_2 is 0.74 Å. Estimate the error in the rotational partition function occasioned by the use of equation 4.5 rather than the more rigorous expression, equation 4.6, at 300 K.

Chapter Five

VIBRATIONAL SPECTRA OF DIATOMIC MOLECULES

The Harmonic Oscillator

If we again look at the infrared spectrum of HCl, but this time at much shorter wavelengths, we find a series of bands of diminishing intensity. The first one occurs at about 2900 cm^{-1} and it can be seen from *Table 5.1* that, like the rotational spectrum, these bands are approximately equally spaced. However, they cannot be rotational lines

Table 5.1 INFRARED BANDS OF HCl

v	$\bar{\upsilon}_{obs}(cm^{-1})$	$\Delta\bar{\upsilon}_{obs}(cm^{-1})$	$\Delta^2\bar{\upsilon}_{obs}(cm^{-1})$
0	(0)		
		2 885.9	
1	2 885.9		103.8
		2 782.1	
2	5 668.0		103.1
		2 679.0	
3	8 347.0		103.3
		2 575.7	
4	10 922.7		

because application of equation 3.6 to these frequencies would suggest an internuclear separation of about 0.1 Å, which is clearly impossible. The energy increments, on the other hand, correspond to the range of energies we would expect for vibrational levels, and for the moment we will assume that these lines are due to transitions between vibrational energy levels, and proceed to establish the spectrum we would expect for a vibrating diatomic molecule.

Assuming the molecule to be in one of its lowest vibrational states, which is always true if we consider absorption at room temperature, we can treat the molecule as a harmonic oscillator to quite a good degree of approximation. We consider again the molecule as a dumb-bell (*Figure 3.1*) whose equilibrium internuclear separation is r_e. If we move either atom from its equilibrium position along the molecular axis, the restoring force is proportional to the displacement, i.e.

$$\text{force} = -k(r - r_e) \tag{5.1}$$

but

$$\text{force} = \text{mass} \times \text{acceleration}$$

so that

$$-k(r - r_e) = m_1 \frac{d^2 r_1}{dt^2} \quad \text{or} \quad m_2 \frac{d^2 r_2}{dt^2} \tag{5.2}$$

If we recall, however, that

$$r_1 = \frac{m_2}{m_1 + m_2} r \quad \text{and} \quad r_2 = \frac{m_1}{m_1 + m_2} r$$

equation 5.2 becomes

$$\begin{aligned} -k(r - r_e) &= \frac{m_1 m_2}{m_1 + m_2} \frac{d^2 r}{dt^2} = \mu \frac{d^2 r}{dt^2} \\ &= \mu \frac{d^2 (r - r_e)}{dt^2} \end{aligned} \tag{5.3}$$

In other words, we can regard the vibrational motion of the diatomic molecule simply as if it were a single mass μ oscillating about its equilibrium position. The potential energy which causes a restoring force of $-k(r - r_e)$ is $V = \frac{1}{2}k(r - r_e)^2$, and, again, it is a fairly easy matter to solve the Schrödinger equation for the motion of a particle of reduced mass μ in this potential field. The solutions of this equation show that the allowed energy levels are

$$\epsilon_v = \frac{h}{2\pi}\sqrt{\frac{k}{\mu}}\,(v + \tfrac{1}{2}) \tag{5.4}$$

in contrast to the classical solution of the harmonic oscillator problem, which is that all energies are possible with a frequency of oscillation

$$\omega = \frac{1}{2\pi}\sqrt{\frac{k}{\mu}} \quad \text{Hz} \tag{5.5}$$

Thus, in quantum mechanics, we have

$$\epsilon_v = h\omega(v + \tfrac{1}{2}) = hc\bar{v}(v + \tfrac{1}{2}) \qquad (5.6)$$

where $\bar{v}$ is in cm^{-1}. In equation 5.6 ω has the dimensions of frequency, but in quantum mechanics it is not to be regarded as a physically meaningful frequency: it is simply a constant which defines the spacing of the energy levels. The constant k is known as the *force constant* of the molecule, and for HCl, where $\bar{v} = 2890\ cm^{-1}$, k is 4.8×10^5 dyne cm^{-1}. We have already seen the form of the wave functions (*Figure 1.2*), and we note again the zero-point energy of the vibrator, which implies that the molecule is always vibrating.

We need the wave functions, of course, to determine the selection rules from the integral $(\int\psi_i \mu \psi_f d\tau)^2$; this integral is only non-zero if (1) $\mu \neq 0$ and (2) $\Delta v = \pm 1$. Thus, to have an infrared spectrum the molecule must have a dipole moment, a rule which is rigorously obeyed, and transitions can take place only one quantum at a time. But we have noted that HCl gives a vibrational spectrum which consists of a series of equally spaced bands of diminishing intensity, in contrast to the single band predicted by the second selection rule (since the transitions $v = 0 \rightarrow 1$, $v = 1 \rightarrow 2$, $v = 2 \rightarrow 3$ should all occur at the same frequency).

The Anharmonic Oscillator

The reason for this discrepancy is that the molecule is not quite a harmonic oscillator, but an anharmonic one (see *Figure 3.9*). There are various possible ways of representing the potential function for the molecule, e.g. as a power series in r

$$V(r) = \tfrac{1}{2} k_1 (r - r_e)^2 + \alpha(r - r_e)^3 + \beta(r - r_e)^4 + \ldots \qquad (5.7)$$

or we can try to find a simple form of potential which has the right type of behaviour but with fewer adjustable constants. There are several simple ones, but easily the most successful and the most useful is that due to Morse. It is usually written in one of two ways, i.e.

$$V(r) = D_e [1 - e^{-a(r - r_e)}]^2 \qquad (5.8)$$

or

$$V(r) = D_e [e^{-2a(r - r_e)} - 2\, e^{-a(r - r_e)}] \qquad (5.9)$$

In equation 5.8 the energy is measured with respect to $V(r) = 0$ at $r = r_e$ and in equation 5.9 with respect to $V(r) = 0$ at $r = \infty$. In both equations D_e is the depth of the potential well (see *Figure 3.9*) and a is an adjustable constant; the effect of changing a is to alter the degree of curvature of the right-hand side of the well. The Morse potential has one minor fault: $V(r)$ should tend to infinity as r decreases to zero, but neither equation 5.8 nor equation 5.9 does this. However, for a molecule such as HCl, the Morse approximation gives a repulsive potential energy of about $+90\,D_e$ at $r = 0$. This value is large enough for the imperfections in the wave functions and selection rules not to be affected to any degree which would be experimentally measurable. This of course does not mean that real molecules are effectively Morse oscillators—but the Morse potential is a very good approximation for some molecules, such as H_2, Cl_2, HCl (see *Figure 5.1*).

The great advantage of the Morse potential over all others is that we can solve the Schrödinger equation with a Morse potential, the allowed energy levels being

$$\epsilon_v = \frac{ah}{2\pi}\sqrt{\frac{2D_e}{\mu}}\,(v + \tfrac{1}{2}) - \frac{a^2 h^2}{8\pi^2 \mu}\,(v + \tfrac{1}{2})^2 \tag{5.10}$$

But

$$\omega = \frac{a}{2\pi}\sqrt{\frac{2D_e}{\mu}}$$

would be the classical frequency of this oscillator, so that by analogy with equation 5.6 we can write

$$\epsilon_v = h\omega(v + \tfrac{1}{2}) - \frac{h^2\omega^2}{4D_e}\,(v + \tfrac{1}{2})^2$$

$$= hc\omega_e(v + \tfrac{1}{2}) - hc\omega_e x_e(v + \tfrac{1}{2})^2 \tag{5.11}$$

where ω_e and $\omega_e x_e$ are in cm^{-1} and

$$x_e = hc\omega_e/4D_e \tag{5.12}$$

The effect of the second term in equation 5.11 is to make the energy levels move closer together as v increases, until the dissociation limit is reached, as can be seen in *Figure 5.1*. The wave functions are no longer symmetrical about $r = r_e$, but are weighted on the side $r > r_e$. *Figure 5.2* shows the probability distribution for the first few levels of HCl, which look like slightly distorted harmonic oscillator distributions. *Figure 5.3* shows the probability distributions for all the vibrational

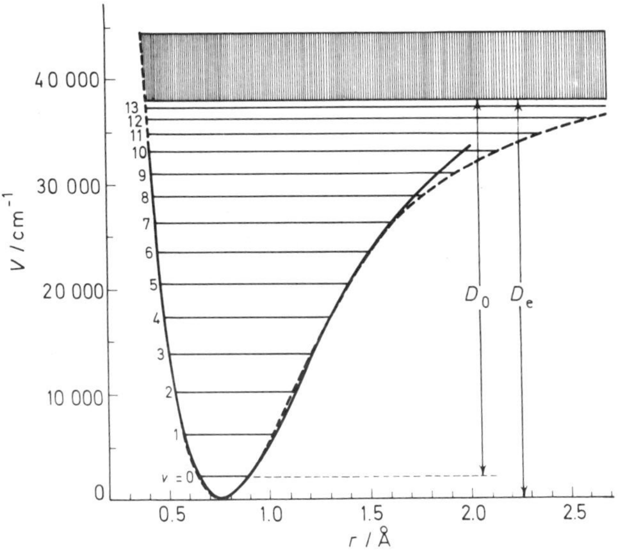

Figure 5.1. Potential curve of the H_2 *ground electronic state with vibrational levels and continuous spectrum. The full curve is drawn according to Rydberg's data. The broken curve is a Morse curve*

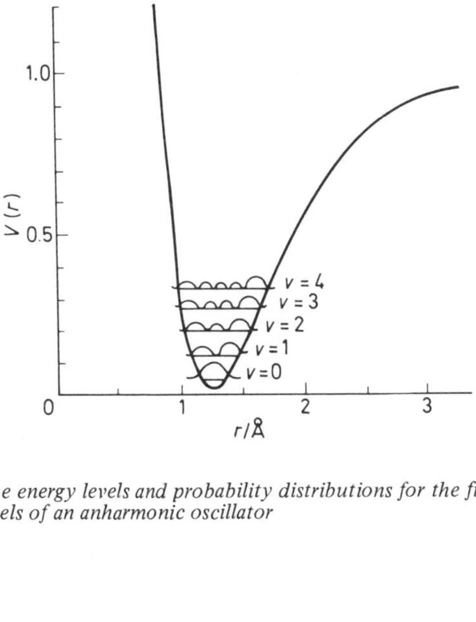

Figure 5.2. The energy levels and probability distributions for the first five vibrational levels of an anharmonic oscillator

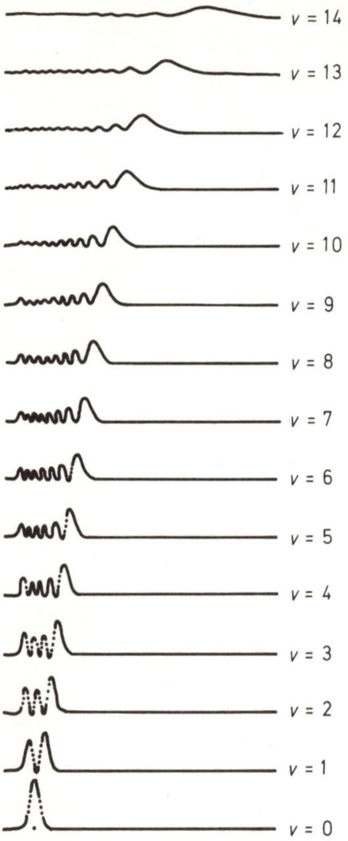

Figure 5.3. Morse wave functions (ψ^2) for the bound state of H_2 *(all functions shown to the same scale). This diagram was constructed from the graphical output of the Manchester University Mercury computer. The functions extend over 5 Å from $\rho = -0.6$ Å to $\rho = +4.4$ Å; the point $\rho = 0$ is marked by a dot below the zeroth-order wave function*

levels of H_2 which are shown in *Figure 5.1* and the effect of the correspondence principle again shows up in the gross loading of the functions towards large values of r at high v—just where on a classical picture the atoms would be moving most slowly and therefore would have the highest probability of being found.

The selection rules, evaluated in the usual way, predict the behaviour observed for real molecules, i.e.

$$(1)\quad \mu \neq 0 \qquad \text{and} \qquad (2)\quad \Delta v = \pm 1, \pm 2, \pm 3, \ldots$$

with the intensities following the pattern shown in *Figure 5.4*, i.e.

$$I_{\Delta v \pm 1} \gg I_{\Delta v \pm 2} > I_{\Delta v \pm 3}$$

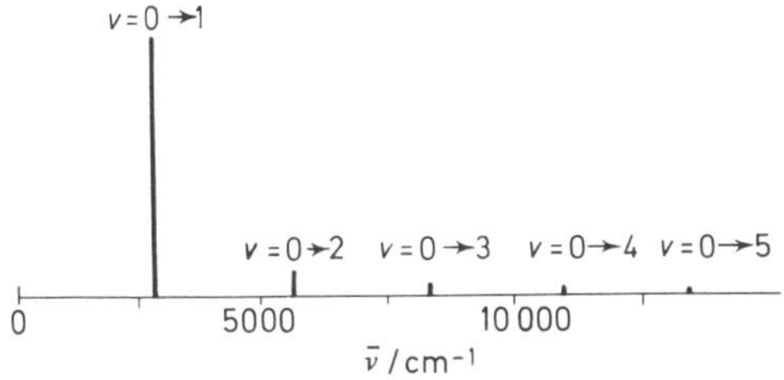

Figure 5.4. Coarse structure of the infrared spectrum of HCl. *The intensity actually falls off more rapidly than indicated by the height of the vertical lines*

Thus, the successive frequencies observed in *Table 5.1* may be simply attributed to the $v = 0 \rightarrow 1$, $v = 0 \rightarrow 2$, $v = 0 \rightarrow 3$, . . . transitions, respectively. The observed spectra of diatomic molecules, in general, lead to an energy level scheme of the form

$$\epsilon_v = hc\omega_e(v + \tfrac{1}{2}) - hc\omega_e x_e(v + \tfrac{1}{2})^2 + hc\omega_e y_e(v + \tfrac{1}{2})^3 - \ldots \quad (5.13)$$

where $\omega_e \gg \omega_e x_e \gg \omega_e y_e$, etc. In practice only the first three terms are within the precision of available spectrometers. (One might add that, while it is not possible to solve the Schrödinger equation with equation 5.7 as a potential, one can show formally that the solutions can be expressed as a power series in $(v + \frac{1}{2})$ such as equation 5.13.)

It is interesting to note that, if the molecule is a Morse oscillator, the *anharmonicity constant* x_e is related to D_e. Usually the fit is not good enough to obtain D_e with any useful accuracy, and it would seem logical, looking at *Figure 5.4,* to try to determine D_e from the convergence limit of the series, equation 5.13 (as one can do for ionisation potentials). Unfortunately, in the infrared it is not possible to do this because of the rapid fall-off in the intensities of the bands. The vibrational spectra of diatomic molecules are of considerable theoretical interest for two reasons: (1) until after the war, when microwave spectroscopy grew up, they were the only means of getting accurate bond lengths through the analysis of the rotational fine structure, and (2) they provide an extremely valuable check on our statistical-mechanical ideas. To understand these two points, however, we must look at our absorption band in a little more detail.

The Rotation–Vibration Spectrum

We have said that the vibrational spectrum of HCl has a band at 2886 cm^{-1} (see *Table 5.1*). If we examine this band under moderate resolution, we find that it consists of two peaks of different intensities (*Figure 5.5*), and that the shapes of these peaks change with temperature. Higher resolution reveals that these peaks are simply the envelope of a series of lines which are approximately equally spaced, apart from a gap in the middle, as can be seen in *Figure 5.6.* The spacing between these lines is close to 20 cm^{-1}, i.e. the same as the line spacing in the pure rotational spectrum. In fact, this band can be completely explained as being due to *simultaneous* transitions between rotational

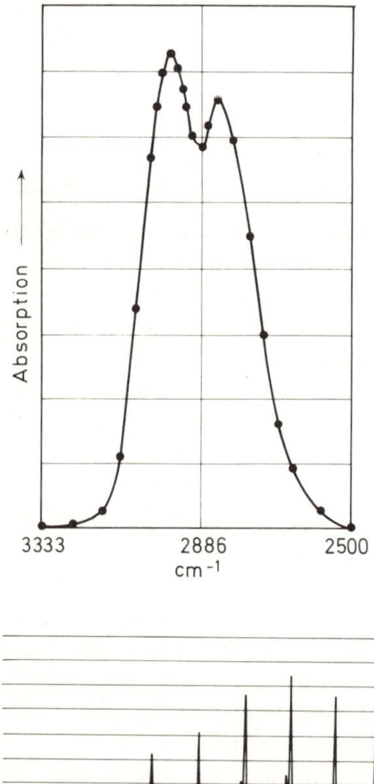

Figure 5.5. The fundamental absorption band of HCl *in the near infra-red under moderate dispersion*

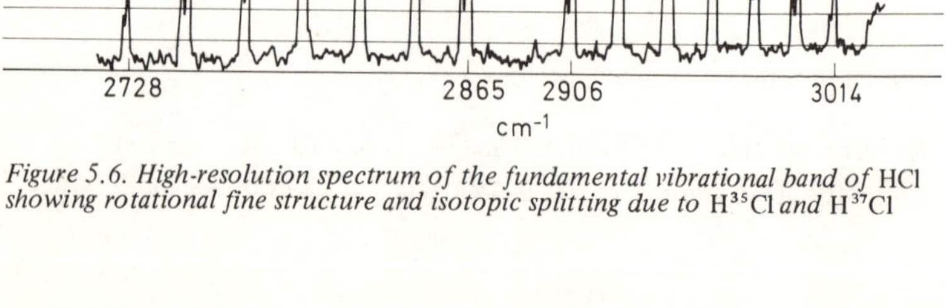

Figure 5.6. High-resolution spectrum of the fundamental vibrational band of HCl *showing rotational fine structure and isotopic splitting due to* $H^{35}Cl$ *and* $H^{37}Cl$

and vibrational energy levels, and for this reason it is known as a rotation–vibration band.

The expected rotation–vibration spectrum is shown in *Figure 5.7*, and it is produced by a consideration of all the possible rotational states of the two vibrational levels concerned, remembering our rotational selection rule $\Delta J = \pm 1$. Previously, in pure rotation spectra *in absorption* we saw only changes of $\Delta J = +1$, but here we must consider also the possibility of $\Delta J = -1$. Thus we would expect two series of equally spaced lines, on either side of the central position (which would correspond to a pure vibrational excitation without any accompanying

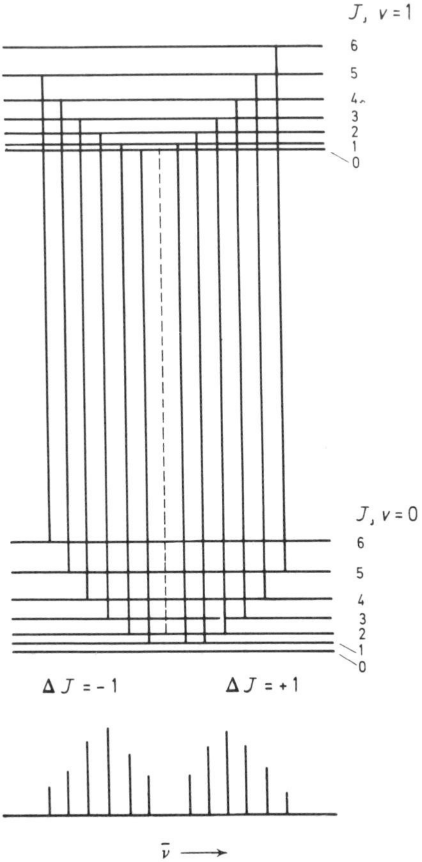

Figure 5.7. Some of the rotation–vibration transitions for a gas-phase diatomic molecule. The absorption spectrum is shown in the lower part of the figure

change in rotational energy). The lines for which $\Delta J = +1$ are called the R *branch* and they lie on the high-frequency (high-energy) side of the centre, because in these transitions the vibrational energy and the rotational energy increase at the same time. Those with $\Delta J = -1$ are the P *branch,* and since a decrease in rotational energy accompanies the transition, these lie on the low-frequency (low-energy) side.

Neglecting for the time being the intensity distribution, a comparison of the predicted and the experimental spectra shows two marked differences:

(1) The central line corresponding to no change in the rotational energy is absent. Therefore $\Delta J \neq 0$ is a selection rule. We cannot infer this by considering either the wave functions of the vibrator or the rotator alone, as we have done previously, and we will accept it as an experimental fact, which can be confirmed if we do the necessary calculation.

(2) The lines are obviously *not* equally spaced. The explanation of this fact is that, as we have already mentioned, the two vibrational states have different bond lengths, i.e. different moments of inertia I, and different rotational constants B. Thus, if r_v is larger in the upper vibrational state, as must always be the case, the transitions for which $\Delta J = +1$ (R branch) will get closer together and those for which $\Delta J = -1$ (P branch) will spread out. This can be seen in *Figure 5.6.* Obviously, then, we can obtain information about the bond lengths in both vibrational states.

We will consider an upwards transition from a vibrational state v'' to a vibrational state v'. Note the convention that $''$ always refers to the lower state in a transition and $'$ to the upper state. In order to describe the transition uniquely, we will need to know the constants given in *Table 5.2.* Then the total energies of the molecule (i.e. the sum of rotation and vibration) in the two levels are

$$\epsilon'' = hc\omega_e(v'' + \tfrac{1}{2}) - hc\omega_e x_e(v'' + \tfrac{1}{2})^2 + \ldots$$
$$+ hcB''J''(J'' + 1) - hcD''J''^2(J'' + 1)^2 + \ldots \qquad (5.14)$$

$$\epsilon' = hc\omega_e(v' + \tfrac{1}{2}) - hc\omega_e x_e(v' + \tfrac{1}{2})^2 + \ldots$$
$$+ hcB'J'(J' + 1) - hcD'J'^2(J' + 1)^2 + \ldots \qquad (5.15)$$

and, of course, the observed frequency is

$$\bar{\upsilon} = \frac{\epsilon' - \epsilon''}{hc} \qquad (3.5)$$

Table 5.2

Lower state	*Upper state*	
v''	v'	Vibrational quantum numbers
J''	J'	Rotational quantum numbers
B''	B'	Rotational constants
D''	D'	Centrifugal distortion constants

We will now consider the frequencies of the lines in the P and R branches separately, and since we are principally interested in the internal structure of the band itself, and the frequency variations within it, we will relate everything to the energy change for the hypothetical transition with $\Delta J = 0$, i.e. to the 'missing line' whose frequency is υ_{vib}. *For the* P *branch* $J' - J'' = -1$, i.e. $J'' = J' + 1$, so that substituting equations 5.14 and 5.15 into equation 3.5 gives

$$\bar{\upsilon}_{\text{P}}^{J'} = \bar{\upsilon}_{\text{vib}} + B'J'(J'+1) - B''(J'+1)(J'+2) - D'J'^2(J'+1)^2 + \\ + D''(J'+1)^2(J'+2)^2 \qquad (5.16)$$

Similarly, *for the* R *branch* $J' - J'' = +1$, i.e. $J'' = J' - 1$ and

$$\bar{\upsilon}_{\text{R}}^{J'} = \bar{\upsilon}_{\text{vib}} + B'J'(J'+1) - B''(J'-1)J' - D'J'^2(J'+1)^2 + \\ + D''(J'-1)^2J'^2 \qquad (5.17)$$

Thus, if we compare the frequency of a pair of lines, with the same value of J' in the upper vibrational state, we find from equations 5.16 and 5.17 that their separation is

$$\Delta\bar{\upsilon}_{\text{P}-\text{R}}^{J'} = (4B'' - 6D'')(J' + \tfrac{1}{2}) - 8D''(J' + \tfrac{1}{2})^3 \qquad (5.18)$$

We could equally well have expressed the frequencies of the P and R branch lines in terms of J'', the rotational quantum number in the lower vibrational state, and from the analogues of equations 5.16 and 5.17 for $\bar{\upsilon}_{\text{P}}^{J}$ and $\bar{\upsilon}_{\text{R}}^{J''}$, we would get

$$\Delta\bar{\upsilon}_{\text{P}-\text{R}}^{J''} = (4B' - 6D')(J'' + \tfrac{1}{2}) - 8D'(J'' + \tfrac{1}{2})^3 \qquad (5.19)$$

We leave the derivation of equations 5.18 and 5.19 as an exercise for the reader.

If we now take the separations represented by equation 5.18 and plot them in the form of $\Delta\bar{\upsilon}_{\text{P}-\text{R}}^{J'}/(J' + \frac{1}{2})$ against $(J' + \frac{1}{2})^2$, we obtain a straight line whose slope is $-8D''$ and whose intercept is $(4B'' - 6D'')$, leading immediately to values of D'' and B'' (hence, r'') for the lower vibrational state. Similarly, taking the separations represented by equation 5.19 and plotting them as $\Delta\bar{\upsilon}^{J''}_{\mathrm{P-R}}/(J''+\frac{1}{2})$ against $(J''+\frac{1}{2})^2$, we likewise obtain D', B', and so r', for the upper vibrational state. The values of B_v obtained in this way for the first six vibrational states of HCl are shown in *Table 5.3*, and it can easily be shown that these values

Table 5.3 ROTATIONAL CONSTANTS OF HCl IN THE DIFFERENT VIBRATIONAL STATES OF THE ELECTRONIC GROUND STATE

v	$B_v(\mathrm{cm}^{-1})$	$\Delta B_v(\mathrm{cm}^{-1})$
0	10.440 0	
		0.303
1	10.136 6	
		0.304
2	9.832 9	
		0.299
3	9.534 3	
		0.302
4	9.232	
		0.299
5	8.933	

correspond to an increase in bond length from r_0 = 1.284 Å to r_5 = 1.357 Å. This table also shows that B_v is approximately linear with v, and can be written in the form

$$B_v = B_e - \alpha_e(v + \tfrac{1}{2}) \tag{5.20}$$

where B_e is the rotational constant of the hypothetical vibrationless molecule, and corresponds to the position represented by the minimum in the potential curve (cf. *Figure 3.9*). Substituting the values from *Table 5.3* in equation 5.20 gives B_e = 10.591 cm^{-1} and α_e = 0.303 (r_e = 1.275 Å). The bond lengths of a great many heteronuclear diatomic molecules, e.g. HF, HCl, HBr, HI, DF, DCl, . . ., CO, NO, . . ., etc., have been obtained in this way.

Isotope Effects

If we look carefully at the spectrum of HCl shown in *Figure 5.6*, each line is seen to consist of a pair of lines which are only partially resolved; they can be completely resolved with a more powerful instrument. This doubling results from the fact that natural HCl is really a mixture of two isotopic species, $H^{35}Cl$ and $H^{37}Cl$, in approximately 3 to 1 ratio.

Consequently, we get a superposition of two rotation–vibration bands, one three times the intensity of the other, each species having its own characteristic spectroscopic constants ω_e, B_v, etc. Thus, a repetition of the analysis of the last section leads to (very slightly) different values of $r_0, r_1, \ldots \omega, D$, etc. for the two isotopic species. The reason that these constants are only very slightly different results from the fact that the potential energy curve $V(r)$ is the same for both molecules, but that μ, the reduced mass (which affects both the vibration and the rotation), is slightly different. A study of equations 3.3 and 5.4 will reveal that the isotope shift in rotational spectra depends on $1/\mu$ whereas the shift in vibrational spectra depends on $1/\sqrt{\mu}$.

A much more marked example of an isotope shift is to be found by comparing DCl with HCl. Regarding the two molecules as harmonic oscillators,

$$\bar{\upsilon}_e(\text{HCl}) = \frac{1}{2\pi c}\sqrt{\frac{k}{\mu_{\text{HCl}}}} \quad \text{and} \quad \bar{\upsilon}_e(\text{DCl}) = \frac{1}{2\pi c}\sqrt{\frac{k}{\mu_{\text{DCl}}}}$$

so that

$$\bar{\upsilon}_e(\text{DCl}) = \bar{\upsilon}_e(\text{HCl})\sqrt{\frac{\mu_{\text{HCl}}}{\mu_{\text{DCl}}}}$$

and taking $\bar{\upsilon}_e$ (HCl) as 2990 cm^{-1}, this gives us $\bar{\upsilon}_e$ (DCl) = 2135 cm^{-1}. In actual fact, the rotation–vibration band of DCl is centred at 2091 cm^{-1} (which can be correctly predicted by a more sophisticated treatment) and of course this spectrum also shows the $^{35}Cl/^{37}Cl$ isotope effect, as before. Two points arise from the comparison of the values of $\bar{\upsilon}_e$ (HCl) and $\bar{\upsilon}_e$ (DCl). First, we can regard it as confirmation that the quantities we have called $\bar{\upsilon}_e$ really do correspond to vibrational changes, because we can explain the magnitude of the HCl/DCl isotope effect. Second, we can show that the two species have markedly different zero-point energies; this effect is even more marked for the pair of molecules H_2 and D_2, where the difference is about 5 kcal mol^{-1}, but it decreases rapidly as the nuclear masses increase.

Population Distribution Among Vibrational Energy Levels

We have already discussed some aspects of this problem, including the calculation of the vibrational partition function. The populations of the various levels fall off as $\exp(-\epsilon_{vib}/kT)$, which is the continuous curve

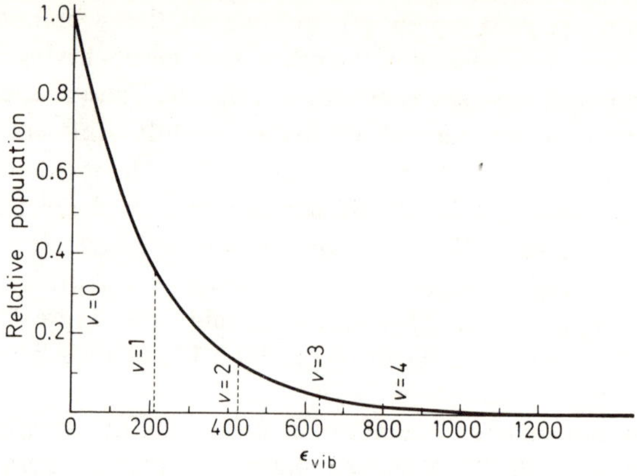

Figure 5.8. Thermal distribution of the vibrational levels. The curve gives the function $\exp(-\epsilon_{vib}/kT)$ *for* $T = 300$ K *with* ϵ_{vib} *in* cm^{-1}. *The broken-line ordinates correspond to the vibrational levels of the* I_2 *molecule*

shown in *Figure 5.8* (for $T = 300$ K). Only I_2, which has a very low value of $\bar{\upsilon}_e$, has an appreciable concentration in the $v = 1$ state at room temperature; the population distribution for this molecule is also shown in the diagram. For most molecules, we have to go to much higher temperatures before we get appreciable population of excited vibrational states. To illustrate this point, the relative populations of the $v = 0$ and $v = 1$ states at 300 K and 1000 K are given for a selection of molecules in *Table 5.4*.

Table 5.4 RATIO OF THE NUMBER OF MOLECULES IN THE *FIRST* TO THAT IN THE *ZEROTH* VIBRATIONAL STATE AT 300 K AND 1000 K

Gas	$\bar{\upsilon}_{0\to1}$ (cm^{-1})	$e^{-(hc\bar{\upsilon}/kT)}$ 300 K	$e^{-(hc\bar{\upsilon}/kT)}$ 1000 K
H_2	4160.2	2.16×10^{-9}	2.51×10^{-3}
HCl	2885.9	9.77×10^{-7}	1.57×10^{-2}
N_2	2330.7	1.40×10^{-5}	3.50×10^{-2}
CO	2143.2	3.43×10^{-4}	4.58×10^{-2}
O_2	1556.4	5.74×10^{-4}	1.07×10^{-1}
S_2	721.6	3.14×10^{-2}	3.54×10^{-1}
Cl_2	556.9	6.92×10^{-2}	4.49×10^{-1}
I_2	213.1	2.60×10^{-1}	7.36×10^{-1}

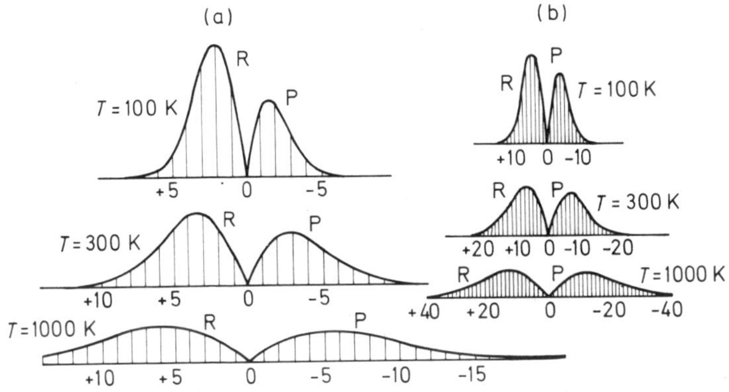

Figure 5.9. Intensity distribution in rotation-vibration bands in absorption at 100 K, *300* K, *and 1000* K *(a) for $B = 10.44$* cm^{-1} (HCl), *(b) for $B = 2$* cm^{-1} (CO)

Intensity Distribution Within the Rotation–Vibration Band

The relative intensities of the rotational lines within the rotation–vibration band are most interesting from a statistical-mechanical point of view. The main factor which governs the relative intensities of the rotational transitions is the relative populations in the initial absorbing states. We have already discussed in Chapter 4 the shape of this population distribution curve, and from our interpretation of the spectrum given in this chapter, we would therefore expect both the P and R branches to have intensity profiles similar to the population distribution curve. This is approximately what is observed (cf. *Figure 5.6*). In actual fact, the two branches of the spectrum are not quite of equal intensity–this is because the intensity does not depend solely on the population, but also to some extent on the value of J. As predicted by the behaviour of the rotational distribution function, if we raise the temperature the over-all absorption area remains roughly constant, but the absorption spreads out and the positions of the maxima move outwards. This effect is shown in detail in *Figure 5.9*.

Exercises

1. The infrared absorption spectrum of CO gas has an intense band centred at 2142 cm^{-1}, the centre line is missing and the spacing in the P

and R branches is 3.8 cm^{-1}. There are similar but much weaker bands at 4256 cm^{-1} and 6342 cm^{-1}, respectively. Calculate the force constant and an anharmonicity constant for the molecule.

2. The fine structure of the fundamental vibrational band of $H^{35}Cl$ contains the following lines

2998.04 cm^{-1}	2863.02 cm^{-1}
2981.00	2841.58
2963.29	2819.56
2944.90	2796.97
2925.90	2773.82
2906.24	2750.13

Estimate r_0 and r_1 and, hence, r_e.

3. From the exact solution of the Schrödinger equation for a Morse oscillator, the allowed energy levels are

$$E = \frac{-D(k - v - \frac{1}{2})^2}{k^2} \qquad v \leqslant (k - \tfrac{1}{2})$$

where D is the depth of the potential energy curve, k is a dimensionless constant characteristic of the molecule and v is the quantum number of the vibration. For the HCl molecule, D = 106 kcal mol^{-1} and k = 23.76. Make, on graph paper, a freehand sketch of a suitable Morse curve and fill in accurately all the vibrational energy levels.

(a) Calculate the relative number of molecules to be found in the $v = 0$, $v = 1$ and $v = 20$ states at 300 K and 2000 K.

(b) Calculate the frequency of the 0-1 vibrational transition.

(c) Show that your energy levels fit the experimental relation

$$\epsilon(v) = hc\omega_e(v + \tfrac{1}{2}) - hc\omega_e x_e(v + \tfrac{1}{2})^2$$

and find ω_e and $\omega_e x_e$.

Chapter Six

ELECTRONIC SPECTRA OF DIATOMIC MOLECULES

Increasing Complexity

Molecular electronic transitions occur in the visible and ultraviolet regions, and are usually responsible for the colours exhibited by chemical substances. They are used extensively in inorganic and organic chemistry for the identification of liquids and solids. We shall, however, only discuss gaseous spectra where we can observe the fine structure of the bands in absorption and in emission.

The major difference between atomic and molecular electronic spectra is that each atomic electronic state is unique, whereas in the molecular case, each electronic state has associated with it both rotational and vibrational degrees of freedom. The situation is illustrated schematically for a pair of electronic states in *Figure 6.1,* and it is clear that a transition between these two states could exhibit a very complicated fine structure.

As in the case of atoms, the electronic states, and the transitions between them, are designated by letters which relate to the orbital angular momentum and the spin angular momentum. For example, an atom is characterised by L (orbital angular momentum quantum number) and S (spin angular momentum quantum number) and a state of some atom might be designated ^1S (i.e. $L = 0, S = 0$), or ^3D (i.e. $L = 2, S = 1$), etc. Likewise, in molecules, we have a spin quantum number which has precisely the same significance, and an orbital angular momentum quantum number which is similar to the atomic quantum number. However, in atoms, the angular momentum vector,

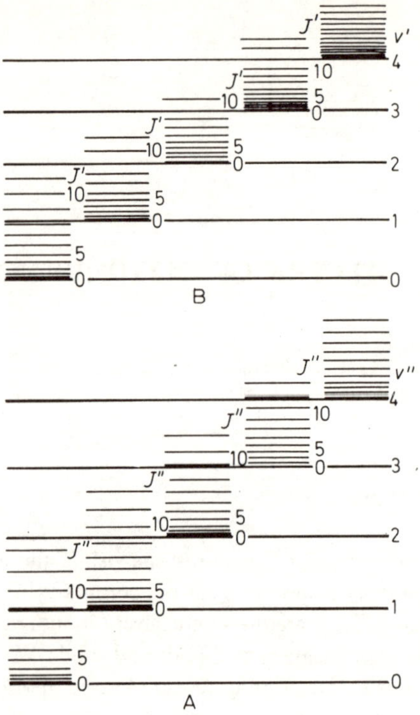

Figure 6.1. Vibrational and rotational levels of two electronic states (A) and (B) of a molecule

defined by L, precesses in one of a number of ways about the applied field and these orientations cannot be defined in the absence of such a field, whereas for a diatomic molecule there is a built-in electric field between the two nuclei, and the orbital angular momentum vector precesses automatically about this axis. The relevant quantum number is called Λ. Corresponding to values of $\Lambda = 0, 1, 2, \ldots$, we have Σ, Π, Δ, . . . states (cf. S, P, D, . . . in atoms), so that a molecule could be designated $^1\Sigma(\Lambda = 0, S = 0)$, $^3\Delta(\Lambda = 2, S = 1)$, etc. Owing to the presence of this internuclear axis, there are some additional symmetry properties associated with molecular wave functions which do not occur in atoms. These are designated by tags, e.g. Σ_g^+, Σ_u^+, Σ^-, Π^-, etc., where g and u apply only to homonuclear molecules; we will not concern ourselves with these symbols, but they may be encountered from time to time. Often, the value of $\Lambda + S$ is used as a subscript, equivalent to the J value in atomic spectroscopy, e.g. $^3\Pi_1$ (instead of say $^3P_1$ as in the Hg atom).

Electronic selection rules are in general of the 'more or less' character. For example, we have already seen that the $6^3P_1 \rightarrow 6^1S_0$ transition of the Hg atom is only 'weakly forbidden' and it is common to use such terms as 'weakly' or 'strongly' forbidden. In general, all electronic transitions are allowed, and the question really is how strong will they be?

The Franck–Condon Principle

Our understanding of the intensity variations in electronic spectra comes through the Franck–Condon principle, which states that when an electronic transition takes place, the internuclear configuration does not change appreciably. We shall see how this applies to a number of simple cases.

The first case is where the minimum in an upper state potential curve is much displaced with respect to the minimum in a lower state potential curve, i.e. $r'_e \gg r''_e$. This situation gives rise to continuous absorption spectra and occurs in F_2, Cl_2, O_2 (and HCl where there is no minimum in the upper curve). The energy of a given transition, and therefore its wavelength or frequency, will depend on the length of the bond in the molecule in question at the time of the transition (cf. *Figure 6.2*). Thus, if the molecule happens to be extended at the instant of absorption, the absorption will take place at longer wavelengths, and if it happens to be compressed, the absorption will be at shorter wavelengths. If we write down the formal expression for the selection rule, we can show that, for any value of r, the intensity of absorption will behave roughly as $|\psi_v \psi_k|^2$, where $|\psi_v|$ is the numerical magnitude of the vibrational wave function in the lower state and $|\psi_k|$ is the numerical value of the vibrational wave function in the upper state, both for the chosen value r. Now, since the upper curve is a smooth repulsion, the value of $|\psi_k|$ *on the potential curve* is almost independent of r, so that the intensity distribution essentially behaves like $|\psi_v|^2$. But this is just the probability distribution of the oscillator, so that the absorption spectrum is most intense where the oscillator spends most of its time, i.e. the absorption curve is essentially a reflection of the ground-state vibrational wave function.

If we now raise the temperature to 1000 K, we will appreciably

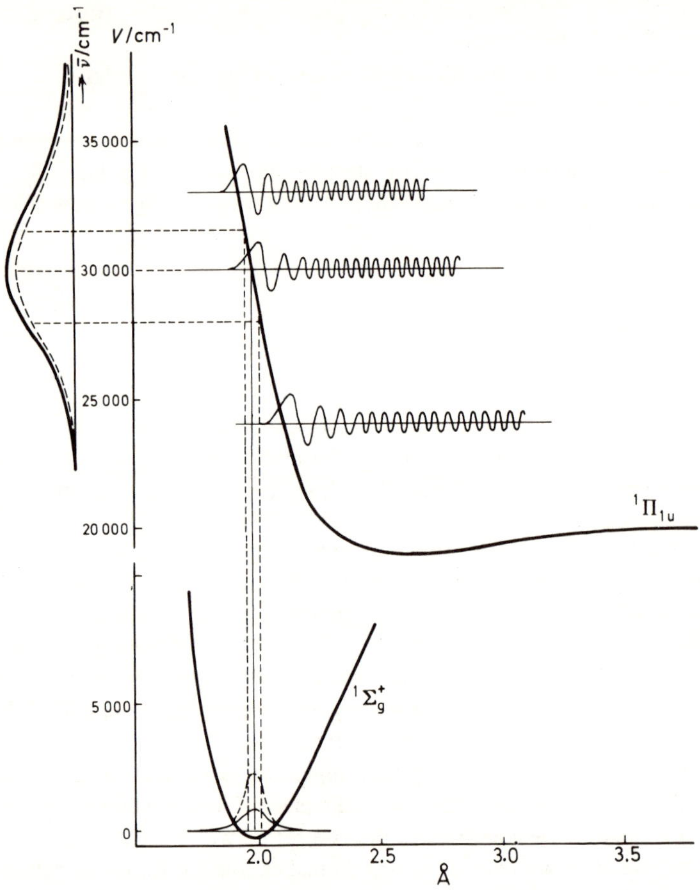

Figure 6.2. Potential curves and repulsive eigenfunctions for Cl_2*. The eigenfunctions are drawn in thin continuous lines. The curve* ψ^2 *for the level* $v'' = 0$ *is shown as a broken line and the reflection of this curve is shown as a broken line which lies under the continuous spectrum*

populate the first ($v'' = 1$) vibrational level, so that transitions will take place from two states whose probability distributions are different. A study of the forms of the relevant probability distributions in *Figure 5.2* shows that as the $v = 1$ state is populated at the expense of the $v = 0$ state the mean distribution curve will flatten slightly, and will spread out a little. Thus, we would expect the spectrum at the higher temperature also to be more spread out, and not quite so intense in the middle, so that the total absorption area remains about the same.

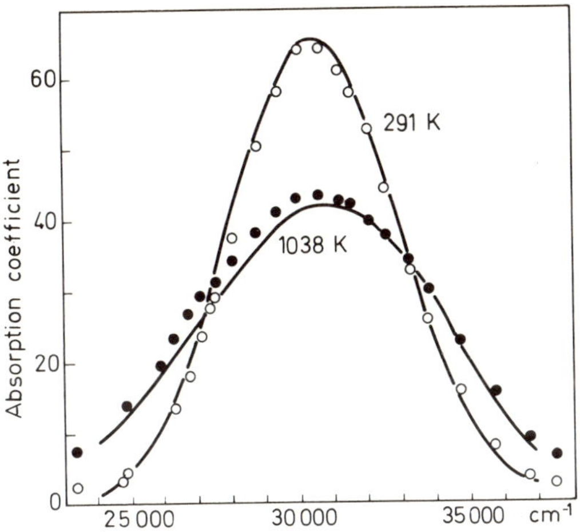

Figure 6.3. Continuous absorption spectrum of Cl_2 *at 291 and 1038* K

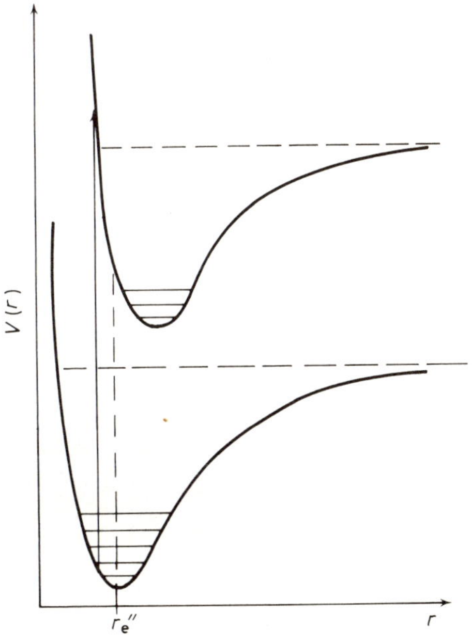

Figure 6.4. Schematic diagram of the ground state and first excited state potential energy curves for Br_2 *and* I_2

Figure 6.3 shows some experimental results for Cl_2 which support these ideas very well.

The second case is where the minimum in an upper state potential curve is moderately displaced with respect to the minimum in a lower state potential curve, i.e. $r'_e > r''_e$. This situation gives rise to a continuous absorption with some fine structure superimposed, as in Br_2 and I_2. The latter is a particularly good example because the upper curve is so positioned that the distance of closest approach in the topmost bound level coincides with r''_e for the lower curve (cf. *Figure 6.4*). This means that on the short-wavelength side (i.e. high-energy) the absorption is analogous to the Cl_2 case, and is continuous. On the low-energy side, however, we have a series of bands due to the transitions $v'' = 0 \rightarrow v'$. There are in fact about 65 of these bands which converge to a limit at 4990 Å, which is roughly at the maximum of the continuous absorption (*Figure 6.5*). The over-all absorption is about the same shape as it would be if the discrete bands were not there, showing

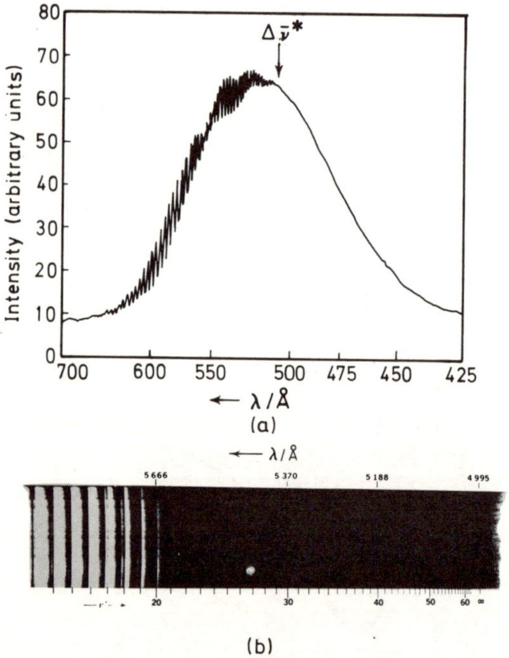

Figure 6.5. Visible absorption band of I_2 vapour showing (a) photoelectric record–the onset of the continuum is indicated at $\Delta\nu^*$; (b) photographic record–the weaker lines based on $v' = 1$ can be seen

that similar considerations govern the intensities in this region also. The bands are not very well resolved in *Figure 6.5(a)*, but under higher resolution it is possible to see clearly the weaker bands due to the series of transitions $v'' = 1 \rightarrow v'$ lying 214 cm^{-1} away from the principal bands; this is because, at room temperature, about one-quarter of the I_2 molecules are in the $v'' = 1$ state. These additional bands, as well as the convergence of the principal series, can be seen in *Figure 6.5(b)*.

It will be noticed that in this spectrum there are very large changes in Δv, but the lines are still quite strong–in contrast with the simple vibrational spectra. Thus, it appears that Δv can be any integer, and the main consideration which governs the intensity is the nature of the electronic transition itself. Notice also that I_2 has no permanent dipole moment, and yet changes in vibrational energy can be seen because they are associated with an electronic transition in which there is a change in the *instantaneous* dipole moment.

Birge–Sponer Extrapolation and Dissociation Energies

From this spectrum of I_2 we can derive much information about the vibrational energy levels of both I_2 and I_2*, despite the fact that neither molecule has a dipole moment. For example, if we look at the spacing between the series based on $v'' = 0$ and $v'' = 1$, respectively, we can get $\bar{v}''_{0\rightarrow 1}$, i.e. an approximate value of ω''_e for the ground state. At the same time, we can, under moderate resolution, observe transitions from $v'' = 0$ (or $v'' = 1$) to most of the v' levels of the upper state and so we have the energy separations between levels of different v'–hence, we can get accurate values of ω'_e and $\omega'_e x'_e$ for this excited state. Furthermore, we can obtain the dissociation energy of this excited state, D'_0 (not to be confused with the centrifugal distortion constant for which we use the same symbol), by carrying out an extrapolation of these vibrational energy levels. We do this by assuming that the vibrational levels get closer together until the spacing becomes zero at the dissociation limit. Thus, if we plot the measured spacings of the levels against v', for all available v', and then extrapolate to zero separation, we obtain a value for the convergence limit Δv^*. If the molecule is a Morse oscillator, such a plot will be a straight line and, in the absence of any evidence to suggest otherwise, we can only make a linear extrapolation on the observed spacings. *Figures 6.6, 6.7* and *6.8* show the form of these plots for H_2, O_2 and K_2, respectively. For H_2, the plot corresponds to the

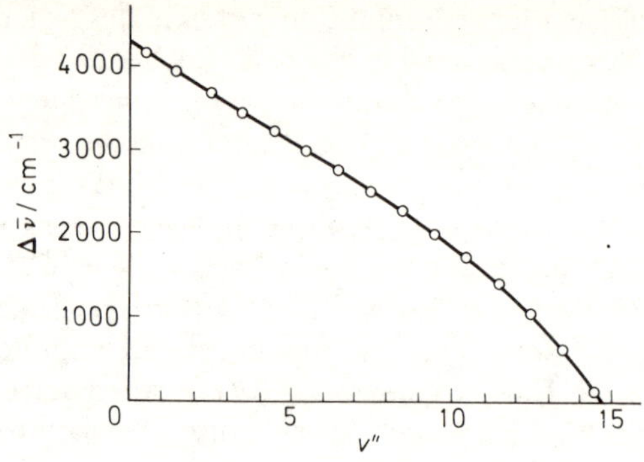

Figure 6.6. Birge–Sponer plot for the ground state of H_2

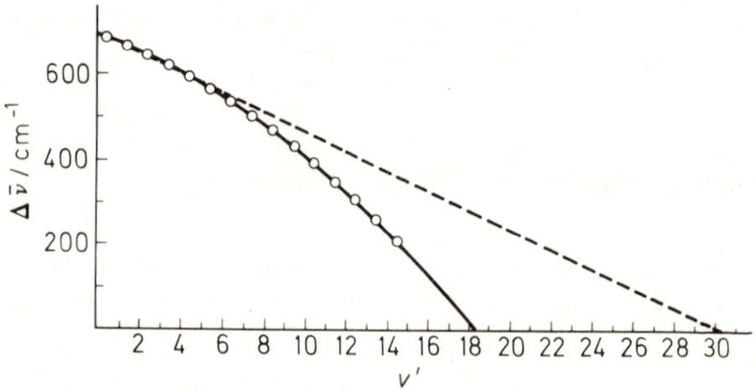

Figure 6.7. Birge–Sponer plot for the excited state $^3\Sigma_u^-$ *of* O_2

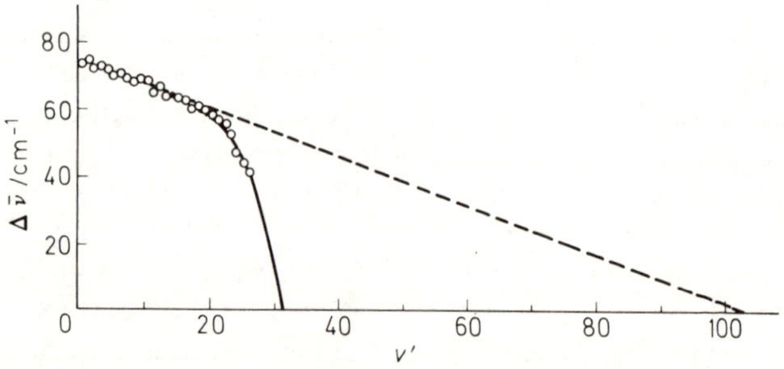

Figure 6.8. Birge–Sponer plot for the excited state $^1\Pi$ *of* K_2

data shown in *Figure 5.1* and confirms that the Morse approximation is a good one for H_2. For O_2, the observed plot is somewhat curved, and the linear extrapolation would predict the existence of a further ten or so vibrational levels. However, for K_2, the deviation from linearity is very marked indeed, and a linear extrapolation grossly overestimates the number of bound levels, and therefore the dissociation energy. For all known molecules the curvature of the plot is always in the same sense, so that a linear extrapolation always overestimates the dissociation energy.

Returning to I_2, where we can observe nearly all the ν' levels, it is a simple matter to extrapolate these to the convergence limit $\Delta\upsilon^*$. If we also know the position of the $\nu' = 0$ level, this gives us directly D'_0, the dissociation energy of the upper state. In the case of I_2, the upper state does not dissociate into two normal $^2P_{\frac{3}{2}}$ iodine atoms, but into one normal atom and one excited $^2P_{\frac{1}{2}}$ atom. But we know very precisely from atomic spectroscopy the energy separation of the doublet $^2P_{\frac{1}{2}} - {}^2P_{\frac{3}{2}}$, and by simply subtracting this value from the energy $\Delta\upsilon^*$ we obtain directly D''_0, the dissociation energy of the ground state. The value derived in this way is in excellent agreement with values from other methods.

The electronic spectra we have discussed so far are special cases which are not very common, and we are left to consider the general case.

In general, the minimum in an upper state potential curve is only slightly displaced with respect to the minimum in a lower state potential curve, i.e. $r'_e \sim r''_e$.

In the first instance, we must realise that any molecule has a large number of excited electronic states, and under low resolution we observe a series of absorptions extending into the far ultraviolet, just as for atoms. These form a Rydberg series converging, as for atoms, to the ionisation limit–hence, we can obtain the ionisation potentials of molecules, i.e. the energies for processes such as $H_2 \rightarrow H_2^+$, $N_2 \rightarrow N_2^+$, etc. However, in the visible or near ultraviolet we only see a few of these absorptions and it is with the fine structure of these that we are now concerned. At room temperature, in absorption, we may see a 'progression'. This results from jumps to various vibrational levels of the upper electronic state from the lowest vibrational level of the ground electronic state (*Figure 6.9*). At higher temperatures, when other vibrational levels become populated, we see other progressions superimposed; I_2 is a good example of this (*Figure 6.5*), although it is just observable in CO. Other examples of progressions are shown in *Figures 6.10* and *6.11*.

In emission, we have already excited the upper electronic state in some way (e.g. heat or electric spark) and we therefore have many

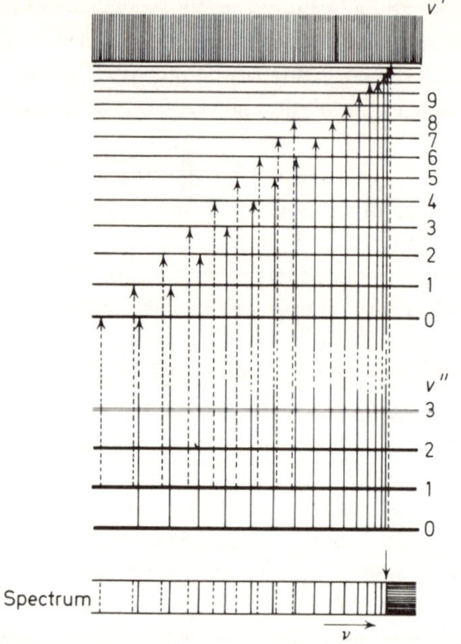

Figure 6.9. Energy-level diagram and schematic drawing of a 'progression' in the vibrational structure of an electronic absorption band. The broken lines represent a progression starting from v'' = 1

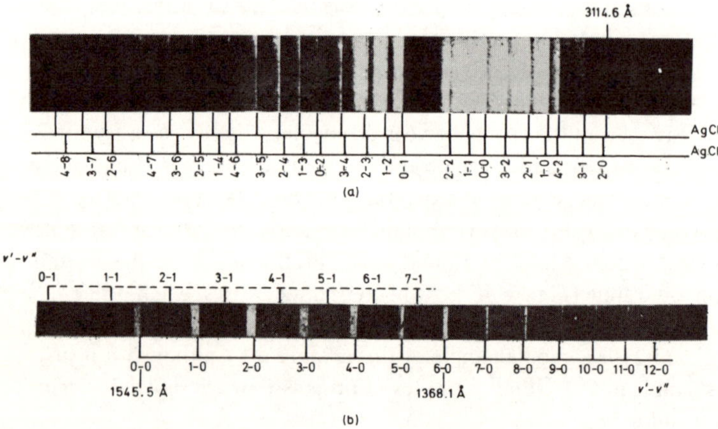

Figure 6.10. Ultraviolet absorption spectra of (a) AgCl *vapour at 900° C and (b)* CO *showing vibrational fine structure*

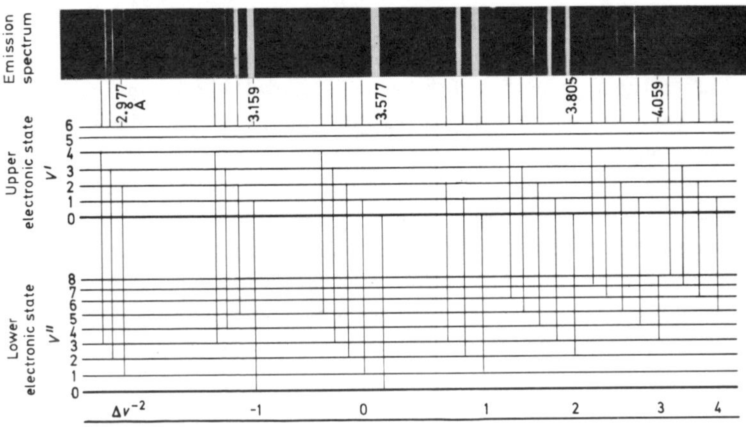

Figure 6.11. Part of the emission spectrum of N_2 *showing several progressions and sequences*

vibrational levels of this state populated. The emission spectrum results from these molecules falling back to lower states, and since there is no restriction in Δv, we see a very large number of associated vibrational transitions. In the case of the molecules N_2, PN and CN, we have a special situation in that the vibrational spacings in the upper and the lower states are almost the same. If the spacings were identical in the two states, we would get just one band for each value of Δv, but in actual fact we observe a set of groups of bands, each group corresponding to a particular value of Δv. Such a spectrum is termed a 'sequence' and *Figure 6.11* shows in detail how it arises.

If we now examine one of the members of this sequence under much higher resolution, we find that it has some fine structure, and that each of the bright lines we saw before is really a 'band head' (*Figure 6.12*).

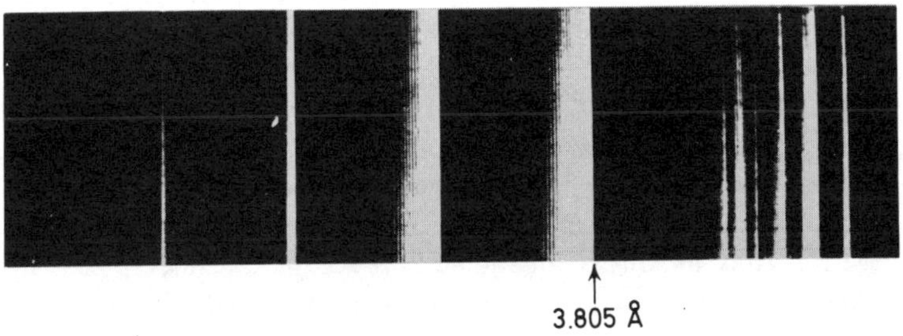

Figure 6.12. Part of an emission band of N_2 *under very high resolution showing rotational fine structure and band heads*

The fine structure is the rotational structure of the bands, and the reason that a band head is formed, as will become clear below, is that the rotational properties of the two states involved are rather more different from each other than was the case with the rotation–vibration spectrum.

If the molecule has $\Lambda = 0$, the rotational selection rule is $\Delta J = \pm 1$, just as it was in the rotation–vibration case. Considering the rotational fine structure of a band, there is, as before, a missing line in the central position, which would correspond to a change in electronic + vibrational energy only, and two branches which we again call P and R. With respect to the frequency υ_0 of this missing line, and neglecting centrifugal distortion, the observed frequencies are

$$\bar{\upsilon} = \bar{\upsilon}_0 + B'J'(J' + 1) - B''J''(J'' + 1) \qquad (6.1)$$

Thus, for the P branch, i.e. $J' = J'' - 1$, we have

$$\bar{\upsilon}_P = \bar{\upsilon}_0 - (B' + B'')J'' + (B' - B'')J''^2 \qquad (6.2)$$

and for the R branch, i.e. $J' = J'' + 1$,

$$\bar{\upsilon}_R = \bar{\upsilon}_0 + (B' + B'')(J'' + 1) + (B' - B'')(J'' + 1)^2 \qquad (6.3)$$

Hence, as from rotation–vibration spectra, we can obtain the bond lengths in either of the vibrational levels concerned by plotting

(i) $\Delta\bar{\upsilon}^{J'}_{P-R}/(J' + \frac{1}{2})$ against $(J' + \frac{1}{2})^2$ giving B''_v

or

(ii) $\Delta\bar{\upsilon}^{J''}_{P-R}/(J'' + \frac{1}{2})$ against $(J'' + \frac{1}{2})^2$ giving B'_v

However, this is not so easy to do because it is much more difficult to pick out the relevant lines. The reason is that B'_v and B''_v differ by much more than they did in the rotation–vibration case. There, they referred to the rotational constants of two different vibrational levels of the *same* electronic state. Now, however, they refer to the rotational constants of two different vibrational levels belonging to *different* electronic states, and as we have already seen, bond lengths can be quite different—although if we have a sequence they cannot be too different. It, as is the general case, $B' < B''$ (i.e. $r' > r''$) the quantity $(B' - B'')$ is negative. Thus, looking at equation 6.2 we see that the lines spread out in the P branch as J'' increases. On the other hand, in the R branch, according to equation 6.3, the lines initially get closer together and, when $(J'' + 1)^2$ becomes large enough, this term takes over and the lines begin to double back on themselves, giving rise to the band head. This is shown diagramatically in *Figures 6.13* and *6.14*. In actual fact, we could get a band head in the rotation–vibration spectrum of, for example, HCl but owing to the similarity between the values of B' and B'' it occurs at such high J values that we cannot observe it.

If the molecule is one in which $\Lambda \neq 0$, we can also have $\Delta J = 0$. This gives rise to an additional series of lines called the Q branch starting at

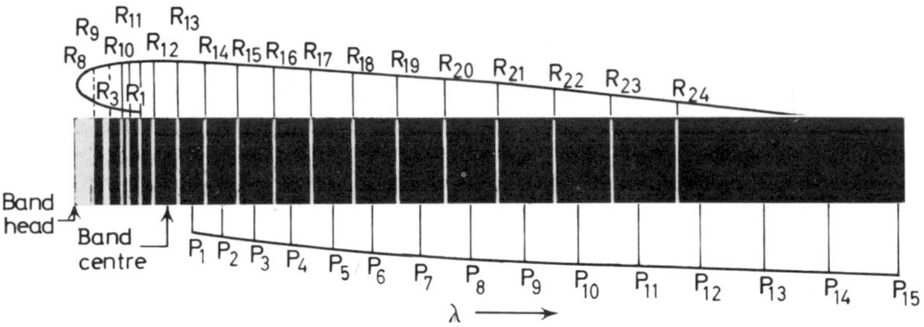

Figure 6.13. Rotational structure of a part of the electronic band of CuH *at 4280* Å

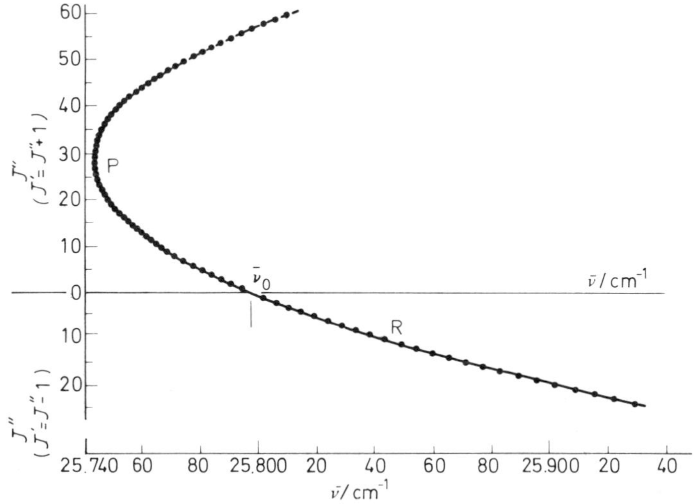

Figure 6.14. Fortrat parabola formed by plotting the frequencies of the rotational lines of the 0–0 component of a CN *band against the rotational quantum number in the lower-energy electronic state. In this case the head is in the* P *branch*

the band origin. Since there is no change in J, the lines are much more closely spaced, the frequencies being

$$\bar{\upsilon}_Q = \bar{\upsilon}_0 + (B' - B'')J + (B' - B'')J^2 \tag{6.4}$$

Generally, $B' < B''$, so that both terms are negative, and the Q branch goes out in the same direction as the P branch. However, since $(B' - B'')$ is numerically smaller than $(B' + B'')$, the spacing is much closer (*Figure 6.15*).

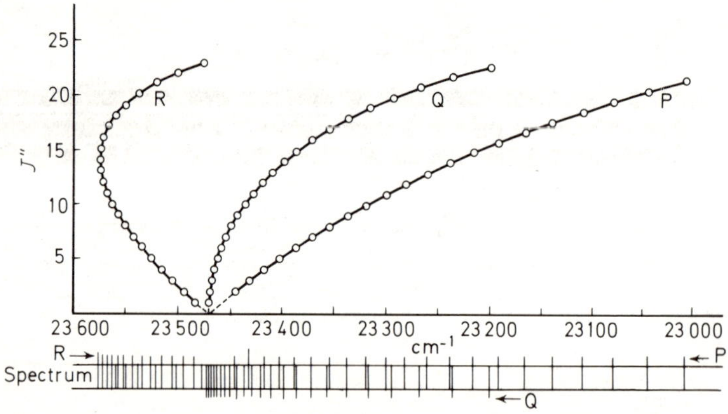

Figure 6.15. Fortrat diagram of the AlH *band. In contrast to Figure 6.14, the Fortrat parabolas of all three branches are drawn above the axis J'' as ordinate. By reflection at the* $\bar{\upsilon}$ *axis, the curve for the P branch would lie in the continuation of the R branch*

Ultraviolet Spectrometers and Spectrographs

In Chapter 3 we outlined the techniques of infrared spectroscopy. Much of the discussion on single- and double-beam spectrometers applies to ultraviolet spectrometers, particularly those operating in the region 4000–2200 Å. Of course the sources and detectors and the materials used for prisms are different, and so are the specifications of the gratings used in the ultraviolet region.

Prisms, Windows

Quartz is the material universally used for prisms, windows and lenses in the ultraviolet region out to 1850 Å. Lithium fluoride has been used out to 1100 Å and calcium fluoride or sapphire out to 1250 Å. Below 1000 Å there is no suitable non-absorbing material from which to construct prisms, lenses, cell windows or lamp housings, and experimental work becomes more difficult.

Sources

Tungsten filament lamps in quartz envelopes are useful sources for the near-ultraviolet region. Discharge lamps operating at about 1000 V (a.c.) and 1 A, containing, for example, 1 mm of hydrogen or helium, emit a continuum useful out to 1000–600 Å.

Detectors

Photographic detection is still a very useful means of obtaining ultraviolet spectra, particularly for high-resolution studies. A variety of photoelectric cells is available for use in ultraviolet spectrometers which produce pen-recorded spectra.

In fact, many of the commercial spectrometers used to obtain the spectra of liquids and solids closely resemble infrared spectrometers, but the instruments used to obtain high-resolution spectra of gases are very different.

Both absorption and emission spectrographs are widely used to record the spectra of gases. The spectrograph uses a grating as the dispersing element and a discharge source for absorption studies or a thermally or electrically excited sample for emission studies. Photographic recording is very common and the whole spectrograph is usually housed in an evacuated tank to allow air to be removed, since O_2, N_2, etc., will absorb at wavelengths $<$2000 Å. Extremely high resolving power is required to separate the rotational fine structure of the electronic bands, even for many diatomic molecules, and consequently spectrographs having focal lengths of several metres are employed.

More Complicated Molecules

We have seen that for diatomic molecules it is possible to derive information about rotation and vibrational constants of both the upper and the lower electronic states in the transition. Thus we can get bond lengths for both states, and dissociation energies (via extrapolation if necessary), for both homonuclear and heteronuclear diatomic molecules, in contrast with infrared spectroscopy. The extension of these techniques to more complicated molecules is now an important and rapidly developing field of spectroscopy. For example, Herzberg has studied the rotational structure of the Rydberg lines for the series of radicals CH_3, CH_2D, CHD_2 and CD_3 and has proved that the radicals are planar. Similarly, Ingold has studied the electronic spectra of acetylene and shown that the upper state in one transition is not linear, but has 120° angles, rather like ethylene without two of its hydrogens. However, much of this work is very difficult owing to the enormous complexity of these spectra when several atoms are present in the molecule.

Exercises

1. The following results were obtained from a study of the vibrational structure of an electronic band at 2000 Å in the emission spectrum of iodine:

v''	(cm^{-1})
0	54 750.9
1	54 536.7
2	54 325.6
3	54 114.8
⋮	⋮
111	42 308.1
112	42 305.5
113	42 304.1
114	42 303.4

These lines represent a progression originating from $v' = 0$ in the upper electronic state:

Calculate for the ground state of iodine:

(a) D_0 and the zero-point energy in cm^{-1};

(b) D_e in kcal mol^{-1};

(c) the frequencies of the first two bands in the vibrational Raman spectrum (see next chapter).

2. Analysis of the fine structure of a band in the ultraviolet spectrum of CO leads to the following equation for the vibrational energy levels in the ground state, for values of v in the range 0–10:

$$E_v = 6200(v + \tfrac{1}{2}) - 36.33(v + \tfrac{1}{2})^2 \text{ cal mol}^{-1}$$

where v is the vibrational quantum number.

Calculate:

(a) a value for the dissociation energy D_e of CO;

(b) the frequency of the fundamental infrared band;

(c) the relative number of molecules in the levels $v = 0$, 1 and 2, at 500 K.

Why would you expect the value of D_e to be an overestimate?

3. The following data were obtained from an analysis of the structure of an electronic absorption band at 1400 Å: a series of approximately equally spaced bands at

64 703 66 231 67 675 69 086 70 470 cm^{-1}

The bands originate from CO molecules in the vibrational state $v'' = 0$. Calculate a force constant for the CO molecule in its excited electronic state and compare the result with that obtained for the ground state in Problem 1, Chapter 5, or Problem 2 above.

4. Explain how it can be that the band head occurs in the P branch in *Figure 6.14*, whereas it occurs in the R branch in both *Figures 6.13* and *6.15*.

5. What is the relationship between the area under a Birge–Sponer plot, for example *Figure 6.6*, and the dissociation energy of the molecule?

Chapter Seven

RAMAN SPECTRA OF DIATOMIC MOLECULES

Only heteronuclear diatomic molecules give rotation and vibration spectra in emission or in absorption in the infrared because of the requirement for a finite, permanent dipole moment. We can, of course, use the fine structure of electronic bands to obtain the vibrational and rotational parameters of a homonuclear diatomic molecule such as O_2 or H_2, but we have seen in the preceding chapter that this is more complex than the analysis of pure rotation or rotation–vibration spectra.

Fortunately, these relatively simple spectra can be observed as frequency shifts in a light-scattering experiment. When a beam of monochromatic visible light is passed through a transparent gas, liquid or solid, some light is scattered with no change in frequency—the Rayleigh effect. This is a weak effect, but if the spectrum of the scattered light is carefully examined, using a spectrometer of high light-gathering power, there is found to be a series of very much weaker lines with frequencies above and below that of the incident beam, i.e. the Raman effect.

The Origin of the Raman Effect

If the frequency of the incident light is ω_i, then its energy is $\epsilon_i = h\omega_i$. The photons which interact with the scattering molecule elastically will retain their energy ϵ_i and consequently their frequency ω_i, resulting in Rayleigh scattering. If, however, the radiation is scattered inelastically,

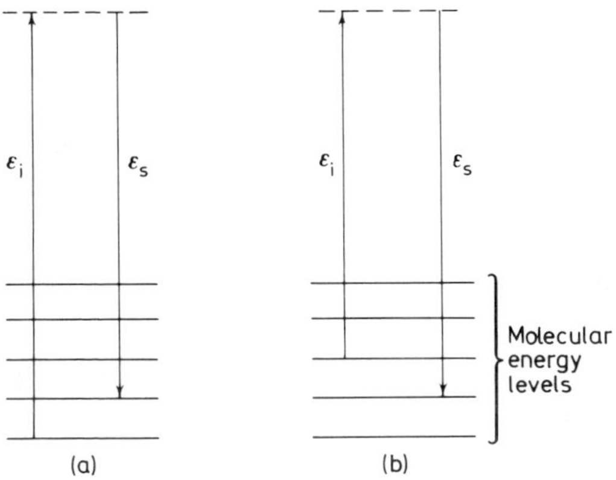

Figure 7.1. Inelastic scattering of light photons by a molecule: (a) the energy of the photon decreases; (b) the energy of the photon increases

the molecule must exchange energy with the photons. Two processes are possible, as shown in *Figure 7.1.*

The molecule may take energy from the photon, thus undergoing a transition to a higher vibrational and/or rotational energy level. (Since visible light is used in Raman spectroscopy, an electronic transition is unlikely to be stimulated.) The radiation is scattered with diminished energy and thus at lower frequency, the difference in energy between the incident and scattered (subscript s) photon corresponding to the energy of the vibrational or rotational molecular transition,

$$\Delta\epsilon = \epsilon_i - \epsilon_s = h(\omega_i - \omega_s) \tag{7.1}$$

i.e.

$$h\Delta\omega = \epsilon_{\text{transition}}$$

Hence, in Raman spectroscopy it is the frequency *shift* from the frequency of the monochromatic incident light beam which is used to calculate the energy of transition, and which corresponds to the frequency of the absorbed radiation in an infrared experiment.

Alternatively, the molecule may be in a higher energy state when it interacts with the light photon and gives up energy to the photon, i.e. $E_s > E_i$, by dropping to a lower molecular energy level. Now

$$\Delta\epsilon = \epsilon_s - \epsilon_i = h(\omega_s - \omega_i)$$

and

$$-h\Delta\omega = \epsilon_{\text{transition}}$$

Raman bands of lower frequency than the incident light beam are called 'Stokes' radiation and correspond to the molecule being promoted to higher energy levels. The radiation scattering at higher frequency is called 'anti-Stokes' radiation. Stokes radiation at a given frequency shift is usually stronger than the corresponding anti-Stokes scattering because at room temperature most molecules are in their lowest energy level, which can give rise only to an upwards molecular transition.

The classical approach is also helpful in understanding the molecular origin of the Raman effect. In a static electric field E, the electron cloud of a molecule is distorted and gives rise to an induced electric moment μ, of magnitude:

$$\mu = \alpha E$$

where α is the polarisability of the molecule. In the electric field due to electromagnetic radiation of frequency ω_i and amplitude E, a fluctuating dipole moment is induced

$$\mu = \alpha E \sin (2\pi\omega_i t) \tag{7.2}$$

and consequently the molecule radiates at the frequency ω_i. This is the mechanism of Rayleigh scattering.

However, the rotational and vibrational degrees of freedom of the molecule cause its components of polarisability, defined with respect to the direction in space of the incident beam, also to vary periodically. The rotational Raman effect arises from the anisotropy of the polarisability of a diatomic molecule, the component of α along the internuclear axis (α_x) being greater than the components ($\alpha_y = \alpha_z$) at right angles. Thus, if ω_r is the rotational frequency,

$$\alpha = \alpha_y + (\alpha_x - \alpha_y) \sin (2\pi\omega_r t) \tag{7.3}$$

Hence

$$\mu = [\alpha_y + (\alpha_x - \alpha_y) \sin (2\pi\omega_r t)] E \sin (2\pi\omega_i t) \tag{7.4}$$

$$= \alpha_y E \sin (2\pi\omega_i t) + \tfrac{1}{2}(\alpha_x - \alpha_y) E [\cos (2\pi [\omega_i - \omega_r] t) - \cos (2\pi [\omega_i + \omega_r] t)] \tag{7.5}$$

The scattered radiation will have frequency components ω_i and $\omega_i \pm \omega_r$, the latter being the Raman scattering. It will only be observed for molecules having anisotropic polarisability ellipsoids, i.e. $(\alpha_x - \alpha_y) \neq 0$.

The vibrational Raman effect is due to the change in the component of polarisability along the internuclear axis during vibration. Suppose this is $\Delta\alpha$; then

$$\alpha = \alpha_0 + \Delta\alpha \sin(2\pi\omega_{vib}t) \tag{7.6}$$

Comparison with equation 7.5 shows that this will result in Raman scattering of frequency $\omega_i \pm \omega_{vib}$.

Thus, to be active in the Raman effect, there must be a change in the molecular polarisability during the vibration or rotation.

Experimental Method

The Raman scattering is approximately 0.01 → 0.001% of the Rayleigh scattering for many simple molecules. Consequently a very strong incident light beam and a very sensitive spectrometer are required. Furthermore, since the Raman bands to be measured lie on either side of the exciting frequency, a monochromatic beam is essential.

Until a few years ago Raman spectrometers consisted of (1) a powerful mercury-arc source, which dissipated several kilowatts of energy, surrounded by filters to select the appropriate line; (2) a sample cell containing about 10 ml of liquid, or several litres of gaseous sample in a multi-pass cell (reflecting the incident beam 10–100 times through the sample); (3) a single monochromator with either a photomultiplier or photographic plate as detector. The scattered radiation is viewed in a direction at right angles to the incident beam. Most work has been done using the λ = 4358 or 5463 Å lines of mercury.

The advent of the laser has considerably improved the situation. The output of a laser is monochromatic and is suitably directed in space in the form of an almost parallel beam. No filters are required and the dimensions of the liquid cell can be considerably reduced, often to a few microlitres. A mirror placed behind the sample cell allows of multiple reflection of the exciting light through the cell.

Typically, 100 mW lasers are used:

helium/neon at	λ = 6328 Å
argon ion at	λ = 4880 Å

A double monochromator reduces stray light from the Rayleigh line and allows of a closer approach to very low frequency shift. A typical layout is shown in *Figure 7.2*. For gases, very small cells can be used

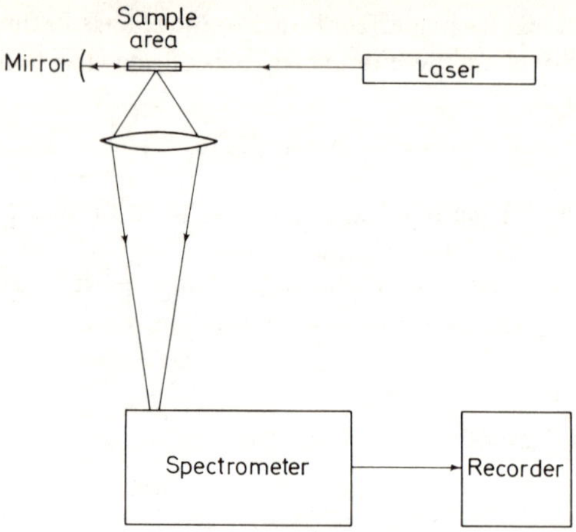

Figure 7.2. Block diagram of a laser-Raman spectrometer

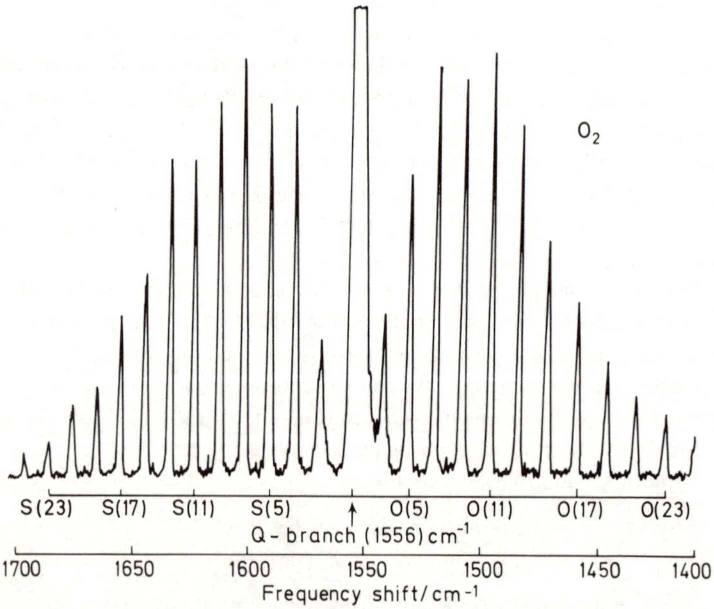

Figure 7.3. Fundamental vibrational band of O_2 *showing rotational fine structure. The Raman spectrometer used an argon ion laser*

with the light focused into the cell. Either photomultipliers or, in the case of the weak emission from gases, photon counting detectors are used. Part of the Raman spectrum of gaseous O_2 is shown in *Figure 7.3*, and the schematic Raman spectrum of HCl vapour is shown in *Figure 7.4* for comparison with the infrared and far infrared spectra discussed

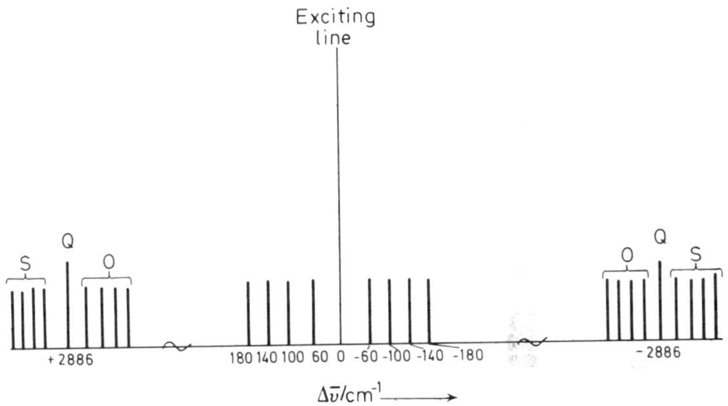

Figure 7.4. Schematic Raman spectrum of HCl *vapour, neglecting centrifugal distortion and assuming $B_0 = B_1$. The pure rotational lines are adjacent to the exciting line. The fundamental vibrational lines are at higher frequency shifts. In practice the intensities of the anti-Stokes lines are greatly reduced*

in Chapters 3 and 5. It will be seen that the small frequency shifts near the Rayleigh line are related to the pure rotational lines observed in the far infrared—except that the spacing is now $\sim 40\ \text{cm}^{-1}$ instead of $\sim 20\ \text{cm}^{-1}$. Vibration–rotation bands are seen at higher frequency shifts, which correspond closely to the frequencies observed in the vibrational infrared spectrum.

Pure Rotational Raman Spectra

All diatomic molecules have anisotropic polarisability ellipsoids and therefore give rotational Raman spectra. The energy levels are given by

$$\epsilon_r = BJ(J + 1) - DJ^2 (J + 1)^2$$

but the selection rules for the changes in rotational level are

$$\Delta J = 0 \quad \text{or} \quad \pm 2$$

and not

$$\Delta J = \pm 1$$

as for far infrared and microwave spectra.

If the molecule is initially in the energy level ϵ'', J'' and is in transition to a new level ϵ', J', then the frequency shift $\Delta\bar{\upsilon}$ of the corresponding rotational line from the exciting line is given by

$$\begin{aligned}\Delta\bar{\upsilon} &= (\epsilon' - \epsilon'')/hc \\ &= B[J'(J'+1) - J''(J''+1)] - D[J'^2(J'+1)^2 - J''^2(J''+1)^2]\end{aligned}$$

If $\Delta J = 0$, $\Delta\bar{\upsilon} = 0$, and this represents Rayleigh scattering.

For $\Delta J = +2$, the scattering lies to low frequency of the incident radiation:

$$-\Delta\bar{\upsilon} = (4B - 6D)(J'' + \tfrac{3}{2}) - 8D(J'' + \tfrac{3}{2})^2 \qquad (7.7)$$

i.e. these are the Stokes rotational lines. The selection rule $\Delta J = -2$ is responsible for the anti-Stokes rotational lines, which also obey equation 7.7 except for a change in sign of the frequency shift.

The structure of the rotational Raman spectrum, typical of a molecule such as HCl or O_2, is shown schematically in *Figure 7.4.* The effect of centrifugal distortion has been neglected in drawing the figure. Note that the first lines on each side of the exciting line are spaced $6B$ from it, and the spacing between rotational lines is $4B$. The rotational constants can be obtained from a plot of $\Delta\bar{\upsilon}/(J'' + \tfrac{3}{2})$ against $(J'' + \tfrac{3}{2})$. The Stokes and anti-Stokes components have comparable intensities. Such spectra are observed, of course, only from gaseous samples.

Vibrational and Rotation–Vibration Raman Spectra

All diatomic molecules have polarisabilities which fluctuate periodically with molecular vibration and therefore all give vibrational Raman spectra The selection rules are identical for the infrared and Raman vibrational spectra,

$$\Delta v = \pm 1, \pm 2, \pm 3, \ldots$$

Thus for a liquid or a solid sample which has no rotational fine structure the fundamental vibrational transition lies on either side of the exciting line at

$$\pm\Delta\bar{\upsilon} = \omega_e(1 - 2x_e) \qquad (7.8)$$

The intensities fall off rapidly and observation of the overtones is very difficult. Their frequency shifts follow the pattern of vibrational overtones in the infrared spectrum. For example, the first overtone lies at

$$\pm \Delta\bar{\upsilon} = 2\omega_e(1 - 3x_e) \tag{7.9}$$

Because the population of the level $\nu = 1$ is much lower than $\nu = 0$ at room temperature, the anti-Stokes vibrational bands are much weaker than the Stokes bands. According to the Boltzmann distribution law, their relative intensities should be

$$\frac{I_{\text{Stokes}}}{I_{\text{anti-Stokes}}} = e^{hc\Delta\bar{\upsilon}/kT} \tag{7.10}$$

With gaseous samples rotational fine structure is observed in vibrational bands. The selection rules are

$$\Delta\nu = \pm 1, \text{etc.}; \qquad \Delta J = 0, \pm 2$$

Thus there is a central line corresponding to a vibrational transition in which the rotational energy level is unchanged. This gives rise to a central line (Q branch) in the fine structure, which is missing in the rotational structure of infrared vibrational bands because in the infrared $\Delta J = 0$ is not allowed. The two branches corresponding to the P and R branches in the infrared spectrum have a spacing of $4B$, since $\Delta J = \pm 2$, and are known as the O and S branches. A particularly fine example of the Stokes vibration–rotation Raman band of O_2 is shown in *Figure 7.3.* Gaseous HCl has a similar band centred at 2886 cm^{-1} with O and S branch spacings of approximately 60, 100, 140, 180, 220, . . . cm^{-1}, from the band centre.

Exercises

1. The bond length of H_2 is 0.74 Å. Its vibrational force constant is 5.3×10^5 dyne cm^{-1}. Draw a schematic diagram of a Raman spectrum of H_2 gas excited by monochromatic light of wavelength 5000 Å.
2. Extend the calculations of Problem 2, Chapter 4 to the relative intensities of the first four rotational Raman lines of HCl.
3. The centre of the vibrational band of HCl vapour lies at 2886 cm^{-1}. Calculate the relative intensities of the Stokes and anti-Stokes components of this band.

Assuming $B_1 = B_0 = 10\ \text{cm}^{-1}$ compare schematically the vibration–rotation structures of the infrared fundamental and the Raman fundamental bands.

Chapter Eight

VIBRATIONAL SPECTRA OF POLYATOMIC MOLECULES

As for diatomic molecules, the vibrational spectra of polyatomic molecules are observed either as Raman spectra or as infrared absorption spectra. The spectra are, however, considerably more complex, but fortunately the basic features can be understood by extending the simple treatment of the harmonic diatomic oscillator given in Chapter 5.

In a diatomic molecule there is one fundamental vibration:

$$\leftarrow \mathrm{O} - \mathrm{O} \rightarrow$$

In a symmetric molecule (homonuclear) this vibration is active only in the Raman spectrum because there is a polarisability change but no change of dipole moment. In a heteronuclear molecule the vibration is both infrared- and Raman-active. For a polyatomic molecule we need to know the number of fundamental vibrations, and their symmetry so that the infrared and/or Raman activity of each vibration can be ascertained.

Number of Fundamental Vibrations

If we consider a polyatomic molecule free to move in space, we can specify the position of *each* atom by three coordinates. Thus if there are N atoms in the molecule, $3N$ coordinates are needed, i.e. there are $3N$ *degrees of freedom.* Since the molecule is free to move *as a whole* but its basic geometry is fixed, we can locate its position by specifying *the three coordinates of its centre of mass.* For a linear molecule *two angles* suffice to identify its orientation in space. Thus five degrees of freedom are required to specify the rotations and translations of the molecule. Therefore, since the only other motion of the molecule is internal vibration, there are $3N - 5$ internal vibration modes, i.e. vibrational degrees of freedom.

The argument for a *non-linear* molecule is similar except that now three angles are needed to identify the orientation of the molecule in space. Thus for a polyatomic non-linear molecule there are $3N - 6$ vibrational modes.

NUMBER OF VIBRATIONAL MODES

N_2O, CO_2	linear	4
H_2O, SO_2	non-linear	3
C_2H_2	linear	7
CH_4	non-linear	9
C_2H_4	non-linear	12
C_6H_6	non-linear	30

Remember that these rules cover only the *fundamental* vibrations. There will be other vibrational bands in the spectrum arising from overtone and combination vibrations, i.e. our argument is equivalent to treating the polyatomic molecule as a set of coupled simple harmonic oscillators.

Normal Modes of Vibration

Consider a bent triatomic molecule, e.g. H_2O, as a set of coupled harmonic oscillators. In vibration, each of the three atoms will execute a set of Lissajous figures with respect to the centre of mass. This is a standard problem in classical mechanics and this complex motion can be represented by the superposition of three simple *normal modes of vibration* which are shown schematically in *Figure 8.1.* The direction and amplitude of oscillation of each atom in each mode are shown by the displacement arrows. The magnitude of the displacement depends on the masses of the atoms and the force constant of the oscillators. Each of the three vibrations has a different fundamental frequency and

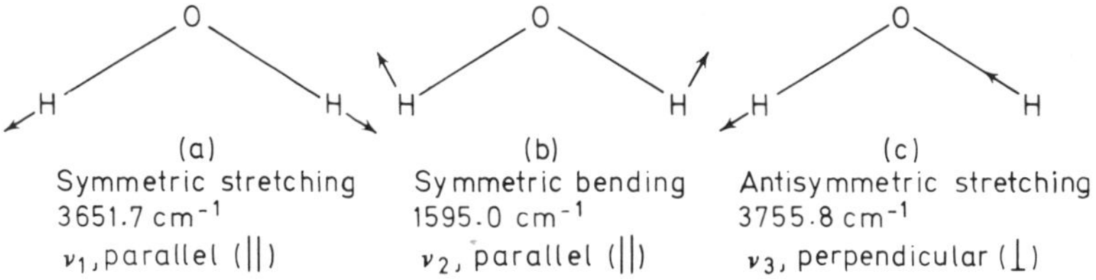

Figure 8.1. Normal modes of vibration of the water molecule

is usually described as a 'stretching' (v_1 and v_3) or 'bending' (v_2) mode, depending on the change in molecular shape.

A *normal mode of vibration* is one in which all the atoms move in phase and with the same frequency. Thus, in a *normal vibration* the atoms all pass through their equilibrium positions at the same time. If we could start with a stationary molecule and displace all the atoms to the extent, and in the direction, of the arrows, the molecule would vibrate indefinitely at the frequency of the chosen normal mode. Of course, we are assuming that the restoring forces are harmonic. In practice they are not, and this leads to the appearance of overtone bands as for diatomic molecules.

If we consider a linear triatomic molecule such as CO_2, we expect four fundamentals, i.e. four normal modes of vibration. These are shown in *Figure 8.2.* However, it will be noted that the two bending

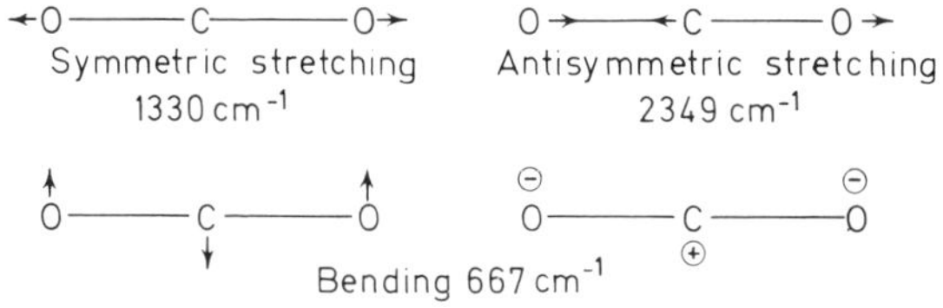

Figure 8.2. Normal modes of vibration of the carbon dioxide molecule. ⊕ and ⊖ signify displacements in and out, respectively, of the plane of the paper

modes are identical except that one is at 90° to the other. Clearly, they will have the same frequency and are said to be 'degenerate' modes of vibration. Thus, only three vibrational frequencies are to be expected for such a molecule.

These arguments can be extended to molecules of considerable size but it becomes more and more difficult to draw the modes as the molecule becomes larger. An attempt to portray the modes for pyramidal and planar molecules of the type XY_3 is made in *Figure 8.3.*

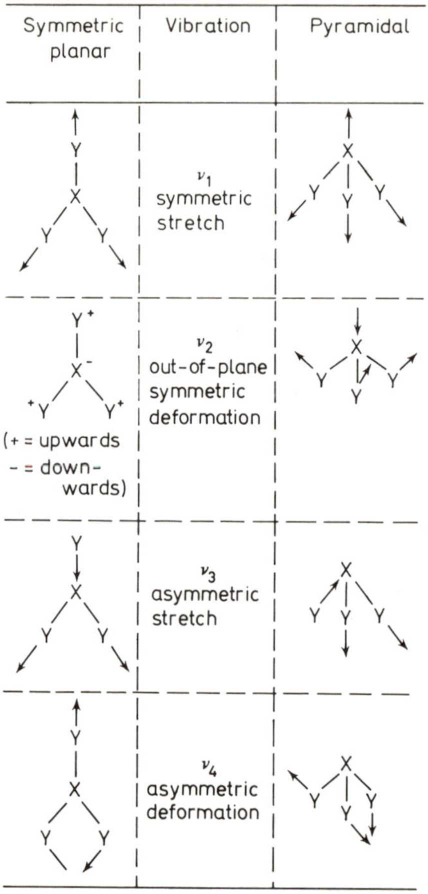

Figure 8.3. Normal modes of vibration of planar and pyramidal XY_3 molecules. Note that υ_3 and υ_4 are each doubly degenerate

Activities of Normal Modes of Vibration

Since there are either $3N - 6$ or $3N - 5$ vibrational modes for a polyatomic molecule, depending on whether it is non-linear or linear, there must be $3N - 6$ or $3N - 5$ sets of energy levels. In the harmonic oscillator approximation these will be given by formulae of the type

$$\epsilon_v = hc\bar{\upsilon}_i(v + \tfrac{1}{2})$$

where υ_i is the frequency of the ith mode. For degenerate modes υ_i will be the same for each vibration in that particular group.

However, not all transitions are allowed. Infrared activity is observed only for those normal modes which produce a dipole moment change during vibration. Raman activity occurs for vibrations which result in a change in polarisability. Consideration of *Figure 8.1* shows that the three fundamental vibrations of H_2O are active in both types of spectra. In the case of CO_2, however, only υ_2 and υ_3 are active in the infrared, whereas υ_1 is active only in the Raman effect.

Symmetry is clearly important in determining the spectral activity of normal modes. All molecules having a centre of symmetry (cf. CO_2) exhibit mutual exclusion, that is to say that all vibrations which are infrared-active do not appear in the Raman spectrum and vice versa.

Now consider the case of H_2O more closely. The molecule has a symmetry axis which bisects the HOH angle and lies in the plane of the molecule. Vibrations υ_1 and υ_2 produce dipole moment changes parallel to this axis, and υ_3 produces a change perpendicular to it. The different symmetry properties are important when considering the rotational fine structure of these vibrational bands.

Not all normal modes of vibration lead to changes in polarisability or dipole moment. Such vibrations are inactive in both infrared and Raman spectra. Two examples, the twisting motion about the C–C bonds in ethane and ethylene, are shown in *Figure 8.4.* The frequency

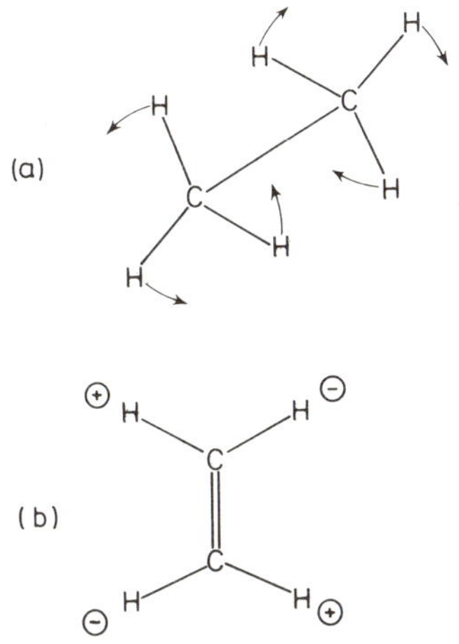

Figure 8.4. Infrared and Raman-inactive torsional vibrations in: (a) ethane; (b) ethylene

of an inactive vibration can be obtained indirectly by determining the frequencies of all the active modes spectroscopically. A statistical calculation then allows a value for the specific heat of the molecule to be calculated, omitting the unknown frequency. Comparison of this value with the observed specific heat gives the contribution (Δc_p) made by the missing frequency υ, since the total specific heat has contributions from all modes of vibration. The inactive frequency can then be calculated, as in Chapter 9, from the equation

$$\Delta c_p = \frac{x^2 e^x}{(e^x - 1)^2} \quad \text{where} \quad x = \frac{h\upsilon}{kT}$$

Overtones and Combination Bands

We might expect from our discussion of diatomic molecules that a polyatomic molecule is really a set of coupled *an*harmonic oscillators. Therefore overtones corresponding to $\Delta v = \pm 2, \pm 3, \ldots$, etc., are observed for each normal mode. As for diatomic molecules, the intensities fall off rapidly. However, a new feature occurs in the spectra of polyatomic molecules, for the selection rules now allow combination bands. They are the combination or difference of two or more fundamental frequencies, e.g.

$$\upsilon_1 + \upsilon_2, \quad \upsilon_1 + 2\upsilon_2, \quad \upsilon_1 - \upsilon_2, \text{etc.}$$

Usually such combination bands are of low intensity.

Vibrational Spectra in Organic Chemistry

A large molecule will obviously have a large number of infrared-active normal modes and it is not surprising therefore that these spectra constitute 'fingerprints' for the identification of organic molecules. The organic chemist is able to recognise frequencies characteristic of group motions, e.g. CH stretch in CH_3, CH_2 or C_6H_5, etc. This information helps to identify the structure of the molecule. In principle, for such a vibrational mode, it is not true that only the C–H atoms are involved in the motion, the mode involves the displacement of most and perhaps all of the atoms in the molecule. However, some motions are approximately separable and thus characteristic of the group concerned. A list of some of these characteristic frequencies is given in *Table 8.1*.

Table 8.1 SOME CHARACTERISTIC GROUP FREQUENCIES IN INFRARED VIBRATIONAL SPECTRA

Group	*Type of vibration*	*Wave number* (cm^{-1})
OH*	Stretch	3600–3700
NH*	Stretch	3350–3500
CH (aromatic)	Stretch	3030
(olefinic)	Stretch	3010–3090
(aliphatic)	Stretch	2840–2970
C≡C	Stretch	2100–2270
C≡N	Stretch	2040–2270
C=O	Stretch	1560–1870
C=C (olefinic)	Stretch	1590–1690
NH	Bend	1510–1640
CH	Bend	1340–1490
CH_3	Bend	1360–1390
OH	Bend	1210–1450
C–O	Stretch	1050–1430
C–N	Stretch	1030–1420
C–F	Stretch	1100–1350
P–F	Stretch	800– 900
P–N	Stretch	700– 740
C–Cl	Stretch	600– 800
C–Br	Stretch	490– 600

* Not hydrogen bonded.

Rotation–Vibration Spectra

Rotational fine structure can only be observed in the spectra of vapours. Polyatomic molecules usually have larger moments of inertia than those of diatomic molecules. Consequently, rotational spacings are smaller than for diatomic molecules and only for the smaller polyatomic molecules can the fine structure be resolved. Usually we see the envelope of all the rotational lines for larger molecules.

The rotational selection rules depend on whether the vibration is accompanied by a change in dipole moment perpendicular or parallel to the symmetry axis.

(a) Linear molecules are the simplest. Here parallel vibrations have selection rules identical with those for diatomic molecules (since the latter can only execute a parallel vibration), i.e.

$$\Delta J = \pm 1 \qquad \Delta v = \pm 1 (\pm 2, \pm 3, \ldots \text{ from anharmonic motion})$$

Thus the bands exhibit P and R branches as shown for the v_1 mode of HCN in *Figure 8.5(a)*. The existence of *only* P and R branches in a fundamental proves that the molecule is *linear.* The linearity of

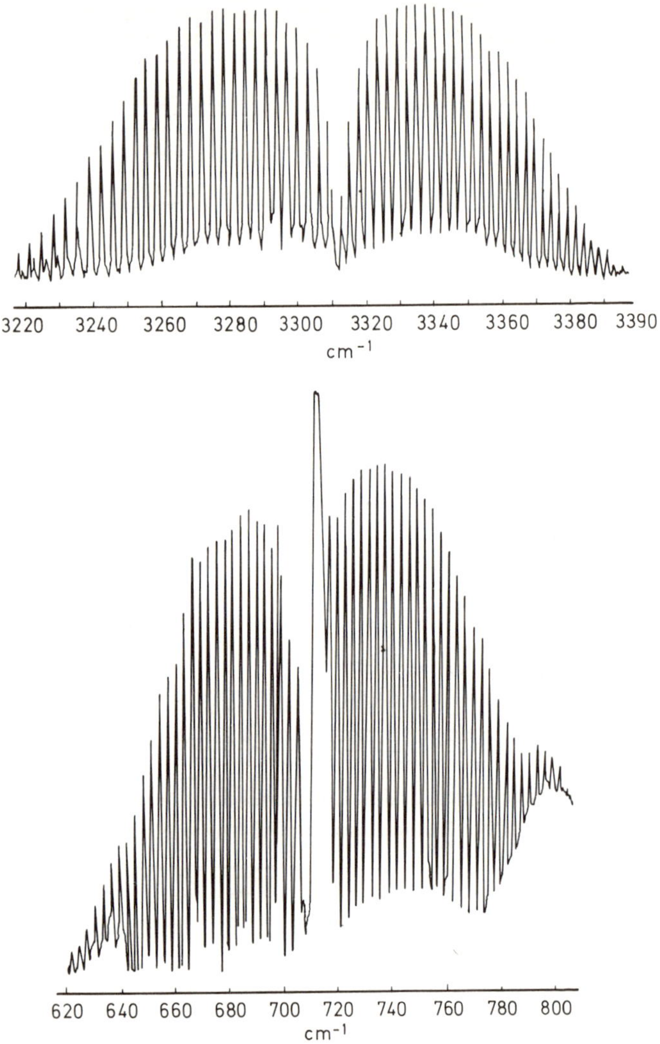

Figure 8.5. Spectrum of (a) parallel symmetric stretch (top) and (b) perpendicular bending (bottom) modes of HCN. *Note the P, R branches in (a) and the P, Q, R branches in (b)*

molecules such as N_2O, HCN, CO_2, C_2H_2 was first established in this way. Perpendicular vibrations have the selection rules

$$\Delta J = 0, \pm 1 \qquad \Delta v = \pm 1 (\pm 2, \pm 3, \ldots)$$

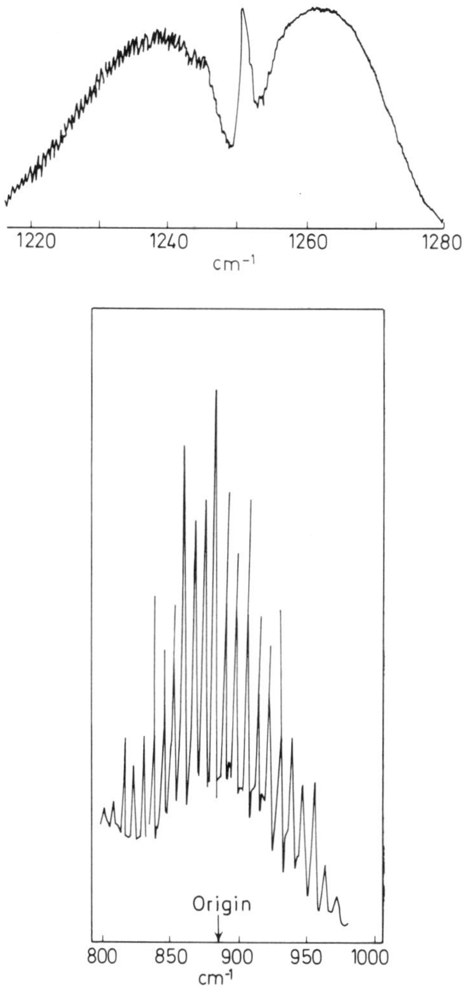

Figure 8.6. (a) Parallel mode (top) and (b) perpendicular mode (bottom) of methyl iodide

which implies that a vibrational change can take place with no simultaneous rotational transition. Thus a Q branch is observed in addition to P and R branches as shown in *Figure 8.5(b)*.

(b) Non-linear polyatomic molecules show fine structure which depends on whether the molecule is a symmetric or asymmetric rotor. As we have noted for the pure rotational spectra, the fine structure is complex, but in all cases $\Delta J = 0, \pm 1$ and thus Q branches will always be observed. *Figure 8.6* shows a parallel and a perpendicular band of CH_3I.

Determination of Molecular Structure from Vibrational Spectra

We have already seen that a linear molecule X–Y–X has only two infrared-active vibrational modes and one Raman-active mode, with different frequencies. On the other hand, the molecules

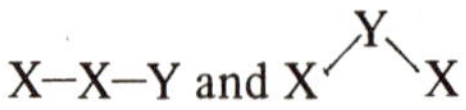

each have the same three frequencies in both the infrared and Raman spectra. Thus, for triatomic molecules we can use vibrational spectra in a simple way partially to diagnose the structures. Mutual exclusion occurs, in fact, in the spectra of any molecule having a centre of symmetry.

If we examine the polarisation of vibrational Raman bands we can derive further information about molecular symmetry. Consider plane polarised light incident along the z-axis of a molecule, with its plane of polarisation in the yz-plane. Suppose the Raman spectrum is viewed along the x-axis and observed with an analyser first in a plane perpendicular to the plane of polarisation and then in a plane parallel to this direction, as shown in *Figure 8.7.* Let the observed intensities of a given Raman band be, respectively, $I_{\parallel}$ and $I_{\perp}$. The *depolarisation* ratio of the vibrational band is defined as

$$\rho = \frac{I_{\perp}}{I_{\parallel}}$$

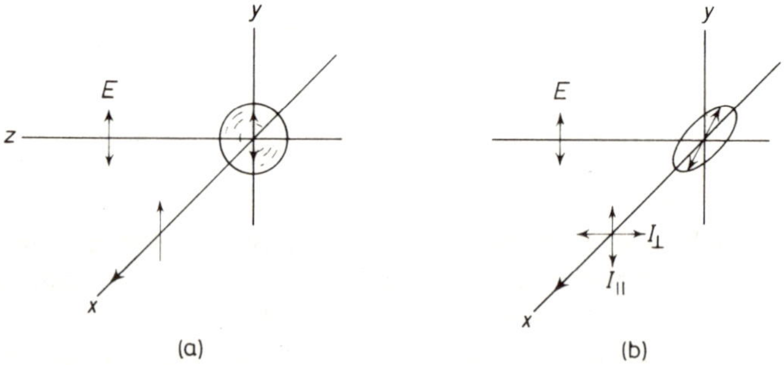

Figure 8.7. Effect of symmetry of vibration on the depolarisation of scattered light: (a) totally symmetric vibration; (b) non-totally symmetric

In a liquid or gaseous sample the molecule will be tumbling or rotating in space, so that the induced electric moment must be averaged over all orientations available to the molecule. For polarised incident light it has been shown theoretically that only totally symmetrical normal modes have $\rho < \frac{3}{4}$. Modes which are not totally symmetrical have $\rho = \frac{3}{4}$. In general, for unpolarised incident light a symmetric vibration has $\rho = 0 \rightarrow < \frac{6}{7}$ and an asymmetric vibration has $\rho = \frac{6}{7}$. Bands having the upper limiting value are said to be 'depolarised' and bands having $\rho < \frac{6}{7}$ are said to be 'polarised'. The symmetrical breathing modes of molecules such as CO_2, CH_4 and C_6H_6 (cf. *Figure 8.8*) have completely polarised Raman lines, i.e. $\rho = 0$. Clearly, knowledge of the degree of polarisation of a Raman line helps to assign it to a particular normal vibration.

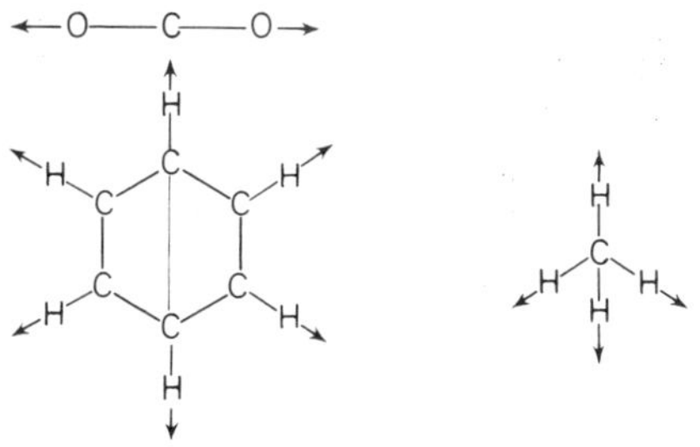

Figure 8.8. Totally symmetric normal modes of vibration

As an example of the use of vibrational spectroscopy in the determination of molecular structure consider the planar and pyramidal XY_3 molecules depicted in *Figure 8.3*. Both have $(3 \times 4 - 6) = 6$ normal modes, but in each case two are doubly degenerate, so that there should be four observable frequencies for each structure. Differences in the predicted spectra occur when we consider the activities of the modes.

Planar XY_3 has:

One mode which is Raman-active only (polarised band).
One mode which is infrared-active only (parallel band).
Two modes which are active in both infrared and Raman spectra (depolarised, Raman, and perpendicular, infrared, bands).

Pyramidal XY_3 has:

All four modes active in both infrared and Raman spectra.
Two frequencies have polarised (Raman) and parallel (infrared) bands, and the other two have depolarised and perpendicular bands.

Thus the structure can be decided on the activities of normal modes alone, though observation of the depolarisation ratios of the Raman bands and the character of the infrared bands is useful corroborative information.

The structure of thionyl chloride, $SOCl_2$, on the other hand, can only be settled by observing the depolarisation ratio of the Raman bands. There are six fundamental frequencies, at 1230, 490, 445, 334, 284 and 194 cm^{-1}. If the molecule were planar there would be three polarised and three depolarised Raman bands. For the pyramidal structure there would be four polarised and two depolarised bands. *Figure 8.9* shows the two components of the Raman spectrum of the

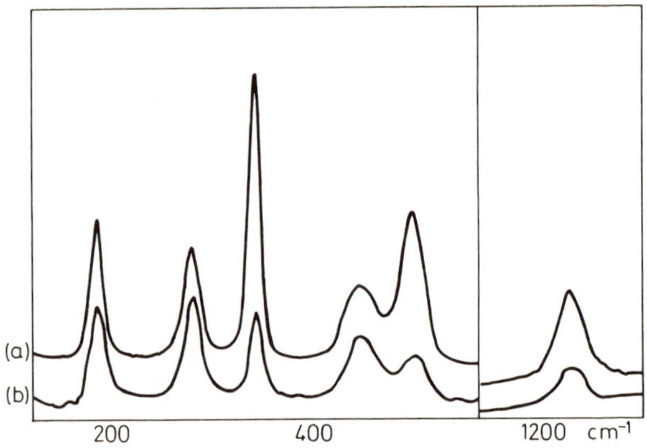

Figure 8.9. Photoelectric record of Raman spectrum of $SOCl_2$. *The upper record (a) was made with parallel polarised incident light, the lower record (b) with perpendicularly polarised incident light. [Note: in the technique used here no analyser was required; instead the plane of polarisation of the incident light was rotated]*

liquid. The bands at 1230, 490, 344 and 192 cm^{-1} are clearly polarised and those at 445 and 284 cm^{-1} are depolarised, which proves that the molecule is non-planar.

One useful feature of vibrational spectroscopy for structure determination is that, along with nuclear magnetic resonance spectroscopy, it can be used for solutions or pure liquids. It can also be used in the vapour phase or on crystalline samples. Electron diffraction methods are restricted to the vapour phase, and X-ray and neutron diffraction methods are useful only on single crystals. *Vibrational analysis, however, is restricted to the determination of the shapes of molecules rather than their sizes,* i.e. bond lengths are not satisfactorily measured by this technique. *Furthermore, the actual shape often can only be fixed with certainty for relatively symmetrical, small molecules, usually with fewer than 12 atoms.*

Chapter Nine

CALCULATION OF THERMODYNAMIC FUNCTIONS FROM SPECTROSCOPIC DATA BY STATISTICAL METHODS

Much of this book (Chapters 2–3 and 5–8) has been concerned with those areas of spectroscopy which can be used to provide the basic molecular parameters needed for the calculation of thermodynamic quantities using the principles outlined in Chapter 1. It should be clearly understood that spectroscopy is an infinitely wider subject, ranging from the very far ultraviolet at wavelengths of a few hundred ångstroms all the way through to nuclear quadrupole spectroscopy at wavelengths of 30 m; the areas of n.m.r. and e.s.r. spectroscopy also have assumed a major importance. The reasons for studying these spectra are many and diverse, and our selection of just those areas which lead to information of thermodynamic interest represents a very biased and restricted account of the subject. We have shown how it is possible by spectroscopic means to determine certain fundamental constants of diatomic molecules by a variety of methods and techniques. Not every method will work for a given molecule—for example, we have seen that homonuclear diatomics do not exhibit infrared rotational or vibrational spectra, but that the required parameters may be found by the use of electronic or Raman spectroscopy—and so on. Our need for thermodynamic information is not, however, confined to diatomic molecules and we have, in various places (and particularly in Chapter 8), indicated how the relevant data may be derived for polyatomic molecules. The remainder of this chapter shows how such information can be used in the calculation of thermodynamic data. The need for accurate thermodynamic parameters extends into many areas of endeavour—for example, in calculating the temperature and composition of rocket exhausts, flames, planetary atmospheres, etc. We cannot measure the required thermodynamic properties in the laboratory at such high temperatures and we have to rely completely on heat capacities, entropies, equilibrium constants, etc., calculated from spectroscopic data by the methods of statistical mechanics.

We have already discussed in Chapters 1 and 4 the calculation of molecular partition functions for translational, rotational and vibrational degrees of freedom, and these results are summarised here.

(i) Translation:

$$f_t = V\left(\frac{2\pi mkT}{h^2}\right)^{\frac{3}{2}} \tag{9.1}$$

(ii) Rotation: using the rigid rotor approximation,

$$f_r = \frac{1}{\sigma}\sum_J (2J+1)e^{-(h^2/8\pi^2 IkT)J(J+1)} \tag{9.2}$$

or when $kT \gg h^2/8\pi^2 I$

$$f_r = \frac{8\pi^2 IkT}{\sigma h^2} \tag{9.2a}$$

(iii) Vibration: using the harmonic oscillator approximation,

$$f_{vib} = \frac{e^{-h\omega/2kT}}{(1-e^{-h\omega/kT})} = \frac{e^{-\epsilon_0^0/kT}}{(1-e^{-h\omega/kT})} \tag{9.3}$$

where ϵ_0^0 is the zero-point energy of the molecule. If the energy of the level $\nu = 0$ is arbitrarily set at zero,

$$f_{vib} = \frac{1}{(1-e^{-h\omega/kT})} \tag{9.3a}$$

For a polyatomic molecule there is one f_{vib} for each normal mode (remembering the degeneracies!).

In most systems at moderate temperatures only the ground electronic level is populated and so the total partition function is, effectively,

$$f = f_t \cdot f_r \cdot f_{vib} \tag{9.4}$$

Thus for N molecules, using the vibrational partition function as in equation 9.3(a),

$$F = -NkT \ln f + E_0^0 \tag{9.5}$$

$$E = NkT^2\left(\frac{\partial \ln f}{\partial T}\right)_V + E_0^0 \quad (9.6)$$

$$S = Nk \ln f + NkT\left(\frac{\partial \ln f}{\partial T}\right)_V \quad (9.7)$$

Other thermodynamic functions can be obtained utilising familiar thermodynamic relations:

(a) $c_V = \left(\frac{\partial E}{\partial T}\right)_V$

(b) $H = E + PV$

$$= E - V\left(\frac{\partial F}{\partial V}\right)_T$$

$$= NkT\left[T\left(\frac{\partial \ln f}{\partial T}\right)_V + V\left(\frac{\partial \ln f}{\partial V}\right)_T\right] + E_0^0 \quad (9.8)$$

and (c) $G = F + PV$

$$= -NkT\left[\ln f - V\left(\frac{\partial \ln f}{\partial V}\right)_T\right] + E_0^0 \quad (9.9)$$

An Ideal Diatomic Gas

Let us consider, for example, the translational, rotational and vibrational contributions to the internal energy of an ideal gas AB:

$$E_t = NkT^2\left[\frac{\partial}{\partial T} \ln\left\{V\left(\frac{2\pi(M_A + M_B)kT}{h^2}\right)^{\frac{3}{2}}\right\}\right]_V = \tfrac{3}{2}NkT$$

Provided that $kT \gg \frac{h^2}{8\pi^2 I}$

$$E_r = NkT^2\left[\frac{\partial}{\partial T}\ln\left(\frac{8\pi^2 IkT}{h^2}\right)\right]_V = NkT$$

$$E_{vib} = NkT^2\left[\frac{\partial}{\partial T}\ln\left(1 - e^{-h\omega/kT}\right)^{-1}\right]_V$$

$$= \frac{Nh\omega\, e^{-h\omega/kT}}{(1 - e^{-h\omega/kT})}$$

When $h\omega < kT$,

$$E_{\text{vib}} = NkT$$

Thus, for one mole of gas, E is $\frac{7}{2}RT$ and, hence, $c_V = \frac{7}{2}R$. Since the molecule has seven degrees of freedom this is consistent with the law of equipartition of energy. It will be noted, however, that the approximations involved in E_r and E_{vib} hold only at sufficiently high temperatures.

From calorimetric measurements we know that at lower temperatures the heat capacity of a diatomic gas varies with temperature in the manner shown in *Figure 9.1*. This variation is readily explained when

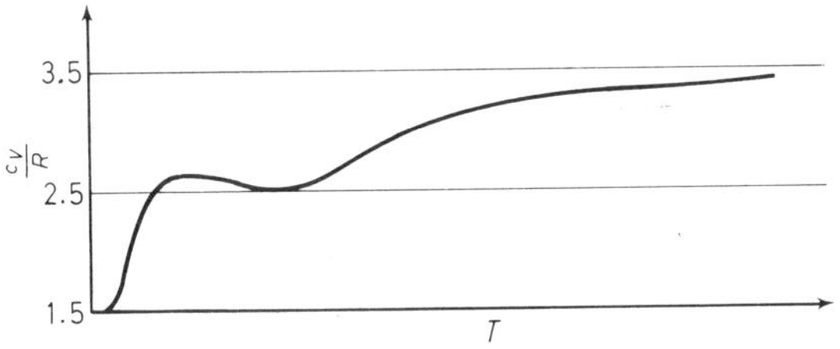

Figure 9.1. Temperature dependence of the heat capacity of a diatomic molecule

we consider the analytical forms of the rotational and vibrational contributions to c_V. The translational energy levels are so closely spaced that at all accessible temperatures $kT > \Delta E_t$ and consequently full equipartition of energy always holds for the translational contribution to energy and heat capacity. However, for the vibrational and rotational energy levels, kT ultimately falls to values less than, or comparable with, their spacings. Consequently equipartition breaks down for these degrees of freedom at lower temperatures.

Rotational Contribution to the Heat Capacities of Gases

In general, at low temperatures, the rotational partition function must be used in the form of equation 9.2. For convenience, let us define a quantity which has the dimensions of temperature:

$$\theta_r = \frac{h^2}{8\pi^2 Ik} \tag{9.10}$$

Then

$$f_r = (2J + 1)\,e^{-(\theta_r/T)J(J+1)} \tag{9.11}$$

Only when θ_r is greater than T can the integrated form of f_r (cf. equation 9.2a) be used. In the present case we must resort to direct numerical computation using equation 9.11 in the expanded form:

$$f_r = 1 + 3e^{-2\theta_r/T} + 5e^{-6\theta_r/T} + 7e^{-12\theta_r/T} + \ldots \qquad (9.12)$$

and calculate the rotational contributions to the heat capacity from

$$c_{V,r} = \frac{\partial}{\partial T}\left[NkT^2\left(\frac{\partial}{\partial T}\ln f_r\right)\right]_V$$

Thus we can construct the curve $c_{V,r}$ as a function of temperature provided we know θ_r, that is, provided we know the moment of inertia of the molecule.

If we write $x = T/\theta_r$, then

$$c_{V,r} = \frac{\partial}{\partial x}\left[Nk\,x^2\left(\frac{\partial}{\partial x}\ln f_r\right)\right]_V \qquad (9.13)$$

We can construct a curve of $c_{V,r}/R$ (for a mole of gas), shown in *Figure 9.2*, which should be common to all molecules and from which

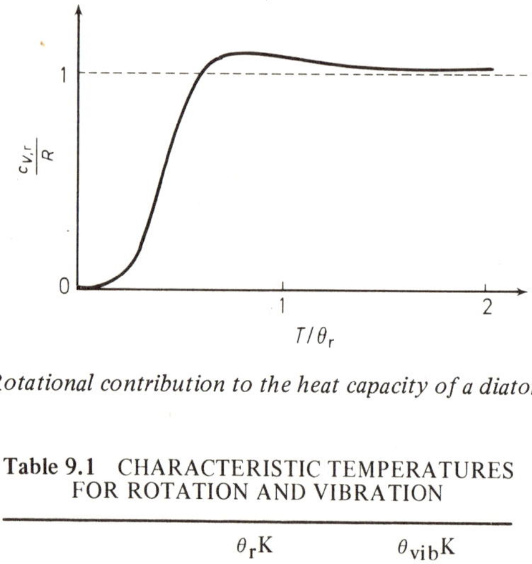

Figure 9.2. Rotational contribution to the heat capacity of a diatomic molecule

Table 9.1 CHARACTERISTIC TEMPERATURES FOR ROTATION AND VIBRATION

	θ_rK	θ_{vib}K
H_2	85.4	6 100
N_2	2.86	3 340
O_2	2.07	2 230
CO	2.77	3 070
NO	2.42	2 690
HCl	15.2	4 140

departures from equipartition can be read off. θ_r is known as the characteristic temperature for rotation and typical values for some common diatomic gases are given in *Table 9.1.* Only when T is greater than θ_r does full equipartition of rotational degrees of freedom occur.

Vibrational Contribution to the Heat Capacities of Gases

It is useful to define also a characteristic temperature for vibration

$$\theta_{vib} = \frac{h\omega}{k} \tag{9.14}$$

thus

$$f_{vib} = \frac{e^{-\theta_{vib}/2T}}{(1 - e^{-\theta_{vib}/T})} \text{ or } \frac{1}{(1 - e^{-\theta_{vib}/T})} \tag{9.15}$$

From equation 9.6

$$E_{vib} = -NkT^2 \left[\frac{\partial}{\partial T} \ln (1 - e^{-\theta_{vib}/T})\right]$$

$$= \frac{NkT^2}{(1 - e^{-\theta_{vib}/T})} \frac{\theta_{vib}\, e^{-\theta_{vib}/T}}{T^2} = \frac{Nk\theta_{vib}\, e^{-\theta_{vib}/T}}{(1 - e^{-\theta_{vib}/T})} \tag{9.16}$$

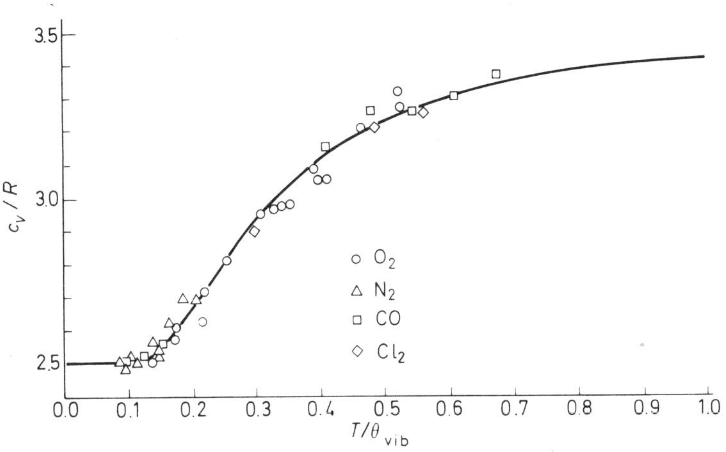

Figure 9.3. Vibrational contribution to the heat capacity of a diatomic molecule. Comparison with experimental results for Cl_2, O_2, N_2 *and* CO. θ_{vib} *is the characteristic temperature for vibration*

Further,

$$c_{V,\text{vib}} = \left(\frac{\partial E_{\text{vib}}}{\partial T}\right)_V = Nk\,\frac{(\theta_{\text{vib}})^2}{T^2}\,\frac{e^{-\theta_{\text{vib}}/T}}{(1 - e^{-\theta_{\text{vib}}/T})^2}$$

$$= Nk\,\frac{(\theta_{\text{vib}})^2}{T^2}\,\frac{e^{\theta_{\text{vib}}/T}}{(e^{\theta_{\text{vib}}/T} - 1)^2} \tag{9.17}$$

Thus, for any diatomic molecule $c_{V,\text{vib}}$ is a common function of θ_{vib}/T. The heat capacity increases with temperature and approaches full equipartition as T exceeds θ_{vib} since $x^2 e^x/(e^x - 1)^2 \to 1$ as x decreases. Values of θ_{vib} for a selection of diatomic molecules are included in *Table 9.1*. The curve is shown in *Figure 9.3* and compared with calorimetric results for several diatomic molecules.

Comparison with Experimental Heat Capacities

In general, the results obtained for the temperature dependence of c_V are in good agreement with experimental data for diatomic molecules. The fall from $\frac{7}{2}R$ to $\frac{5}{2}R$ has been observed for many gases but only for H_2, HD and D_2 have measurements been extended to temperatures which were sufficiently low to observe the fall to $\frac{3}{2}R$. However, three common diatomic gases, H_2, O_2 and NO, show unexpected anomalies.

H_2 : The experimental heat capacity curve for H_2 in the neighbourhood of θ_r is shown in *Figure 9.4*. It bears no resemblance to the curve calculated from equation 9.13. Normal H_2 is a simple mixture of two

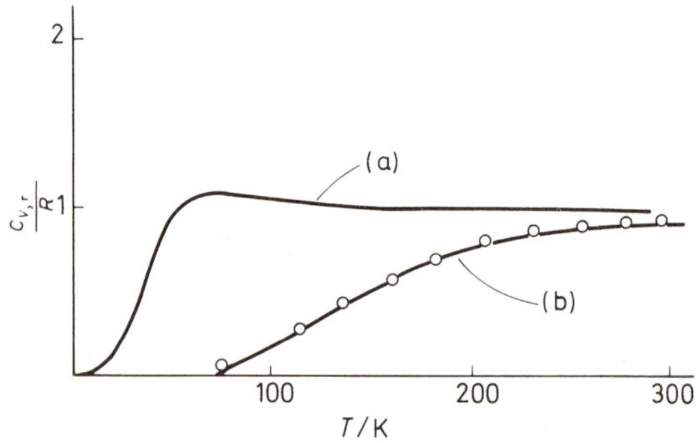

Figure 9.4. Heat capacity of H_2. ○—*experimental; curve (a) from equation 9.11, curve (b) from combination of equations 9.18 and 9.19*

types of molecule, *ortho* and *para,* which arise from the fact that the proton has a nuclear spin quantum number of $\frac{1}{2}$. Thus *ortho*-H_2 molecules with parallel spins have antisymmetrical rotational wave functions (J is odd), whereas the *para*-H_2 molecules with antiparallel spins have symmetric rotational wave functions (J even); this is because the Pauli Exclusion Principle requires that the *total* wave function of the molecule be antisymmetric. At high temperatures, under the usual conditions for the formation of H_2 the equilibrium proportions of *ortho*/*para* is 3 : 1. On cooling, this equilibrium ratio is preserved because the rate of interconversion between the two species is extremely low. The rotational partition function of *ortho*-H_2 is derived from the sum of energy states with odd values of J:

$$f_{\mathrm{r}}^{ortho} = 3\sum_{J=1,3,5} (2J+1)\,\mathrm{e}^{-J(J+1)\theta_{\mathrm{r}}/T} \tag{9.18}$$

whereas the rotational partition function of *para*-H_2 is obtained by summing the states with even J values:

$$f_{\mathrm{r}}^{para} = \sum_{J=0,2,4} (2J+1)\mathrm{e}^{-J(J+1)\theta_{\mathrm{r}}/T} \tag{9.19}$$

The heat capacity curve calculated for a 3 : 1 mixture is shown in *Figure 9.4* to be in good agreement with experiment.

Other homonuclear molecules in which the atom has spin angular momentum (e.g. D_2) will show similar departures from the rotational heat capacity curve calculated from equation 9.13 and drawn in *Figure 9.2.* The heteronuclear molecules, including HD, should be in accord with *Figure 9.2.*

O_2 and NO: The anomalous behaviour of O_2 and NO arises from an entirely different contribution, due to the fact that the two lowest electronic energy levels lie very close together. In most molecules the gap between the ground state and the first electronic excited state is so large that the thermal excitation is not appreciable at accessible temperatures. Thus the electronic contribution was assumed to be unity in equation 9.4. O_2 has its lowest electronic states separated by 22.7 kcal mol^{-1}. In NO the ground state is part of a multiplet of levels in which the next lowest lies at 0.35 kcal mol^{-1}.

Both molecules can be considered to have an electronic partition function arising from a two-level energy system. If we denote the lowest level by energy ϵ_1 and degeneracy g_1 and the second level by ϵ_2 and g_2, then the electronic partition function is

$$f_{\mathrm{e}} = g_1\mathrm{e}^{-\epsilon_1/kT} + g_2\mathrm{e}^{-\epsilon_2/kT} \tag{9.20}$$

If we choose the energy of the molecule in the ground state arbitrarily as zero, then

$$\Delta\epsilon = \epsilon_2 - \epsilon_1$$

and

$$f_e = g_1 + g_2\, e^{-\Delta\epsilon/kT}$$

Thus

$$\epsilon_e = NkT^2 \left[\frac{\partial}{\partial T} \ln\,(1 + g\, e^{-\Delta\epsilon/kT}) \right]$$

$$= \frac{N.\, \Delta\epsilon\, .\, e^{-\Delta\epsilon/kT}}{(1 + g\, e^{-\Delta\epsilon/kT})} \qquad (9.21)$$

where $g = g_2/g_1$, and

$$c_{V,e} = Nk \frac{(\Delta\epsilon)^2}{kT} \quad \frac{g e^{-\Delta\epsilon/kT}}{(1 + g\, e^{-\Delta\epsilon/kT})^2} \qquad (9.22)$$

Plotted as a function of $(kT)/(\Delta\epsilon)$, this curve is shown in *Figure 9.5*; the peak is close to $(kT)/(\Delta\epsilon) = \frac{1}{2}$ and the height of the curve depends on g.

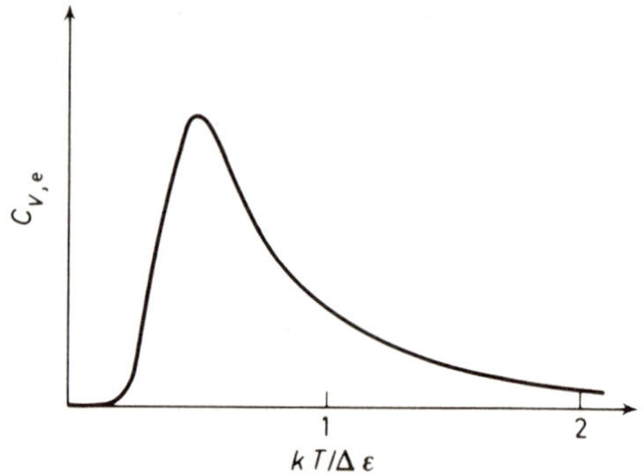

Figure 9.5. Electronic contribution to heat capacity of a molecule with a low-lying excited electronic state. $\Delta\epsilon$ is the separation between the ground state and the excited state

Note that at higher temperatures $c_{V,e}$ falls to zero because the two levels become effectively equally populated and energy can no longer be taken up by a net transfer of molecules to the upper state.

For NO, $g = 1$, and the calculated peak of the electronic contribution is about $0.44R$. This can be compared with the experimental data given in *Figure 9.6.* The peak occurs at 170K for NO, but at 1000K for O_2, which emphasises the larger value of ΔE in O_2.

The very distinctive individual contributions of rotational, vibrational and, for O_2 and NO, electronic degrees of freedom to the total heat capacity of a diatomic molecule are best seen by comparing *Figures 9.5, 9.3* and *9.1.*

The agreement between calculated and experimental heat capacities for polyatomic molecules is equally satisfactory, as is shown in *Table 9.2.* For each molecule, there is a common contribution of $5R/2$ to the heat capacity at constant pressure from the translational motion, as shown in column 3. The number of rotational degrees of freedom, as we have already noted, depends upon whether or not the molecule is linear, and this reflects itself in the difference between a rotational contribution to the heat capacity of R for acetylene and $3R/2$ for

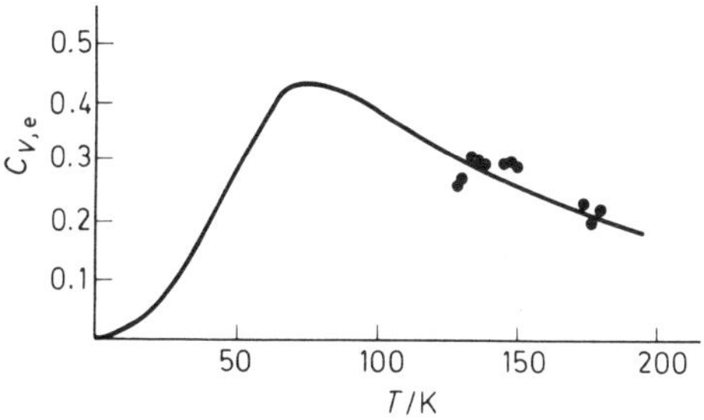

Figure 9.6. Theoretical curve and experimental results (•) for the electronic contribution to the heat capacity of NO

Table 9.2 HEAT CAPACITIES OF POLYATOMIC MOLECULES IN cal deg^{-1} mol^{-1}

Molecule	*T*(K)	$c_{p,t}$	$c_{p,r}$	$c_{p,vib}$	*Total* c_p	
					Calculated	*Observed*
CH_4	298	4.97	2.98	0.57	8.52	8.57
	481	4.97	2.98	2.87	10.82	11.20
	1 000	4.97	2.98	9.22	17.17	–
C_2H_2	288	4.97	1.99	3.39	10.35	9.97
	500	4.97	1.99	6.12	13.08	–
$CH_3C{\equiv}CH$	272	4.97	2.98	5.82	13.77	13.76
	369	4.97	2.98	8.61	16.56	16.52
Cyclo-C_3H_6	272	4.97	2.98	4.18	12.13	12.10
	368	4.97	2.98	8.87	16.82	16.77
$CH_3C{\equiv}CCH_3$	336	4.97	3.97	11.20	20.14	20.21

methane. The vibrational contribution increases with temperature for every molecule, is expected, and also increases roughly with increasing numbers of vibrational degrees of freedom. The final entry, dimethyl acetylene, is an example of a special case—here there is an extra rotational degree of freedom, since one methyl group can rotate freely about the C–C–C–C axis with respect to the other methyl group. Hence, the rotational contribution in this case is $2R$ instead of $3R/2$. In molecules such as ethylene and ethane, we have an intermediate situation where the internal rotation is more or less restricted, but a discussion of this aspect of the problem is beyond the scope of this book. Finally, we call attention again to the fact that some capacities are listed in *Table 9.2* for high temperatures at which, as yet, measurements have not been made.

Equilibrium Constants

By use of equation 9.9, the Gibbs free energy of a gaseous substance can be calculated. Hence, if the relevant molecular parameters and spectroscopic data are available for all the reactants and products in a gaseous chemical reaction, the standard free energy change can be calculated. Then from the relation

$$\Delta G^0 = -RT \ln K \tag{9.23}$$

we can evaluate the equilibrium constant.

Equation 9.9 can be simplified if we assume the gas to be ideal, since

$$\left(\frac{\partial \ln f_r}{\partial V}\right)_T = \left(\frac{\partial \ln f_{vib}}{\partial V}\right)_T = 0$$

and

$$\left(\frac{\partial \ln f_t}{\partial V}\right)_T = \frac{1}{V}$$

Thus equation 9.9 becomes

$$G = -NkT(\ln f - 1) + E_0^0 \tag{9.24}$$

and for 1 mol

$$G^0 = -RT \ln f_t f_r f_{vib} + E_0^0 + RT \tag{9.25}$$

Consider now a typical reaction, say

$$A + B \rightarrow C$$

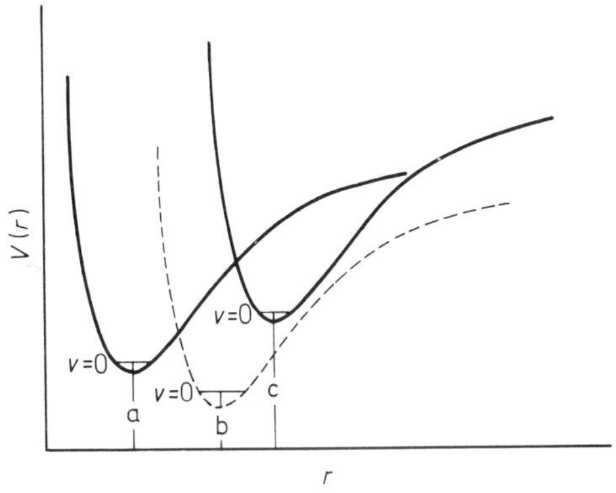

Figure 9.7. Schematic potential energy curves for species A, B and C

When calculating ΔG^0, we must remember that the partition function for each species involved in the reaction is evaluated relative to a different reference point; thus a common zero must be chosen. If we consider the reaction to occur at 0 K, then the molecules have zero energy of translation and rotation and only the zero-point vibrational energy. Their potential energy curves may lie at different levels on an absolute energy scale, as shown diagramatically for three diatomic molecules, A, B and C, in *Figure 9.7.* Let us choose the arbitrary zero shown in the figure. Then

$$G^0_{\mathrm{A}} = -RT \ln \; f^{\mathrm{A}}_{\mathrm{t}} f^{\mathrm{A}}_{\mathrm{r}} f^{\mathrm{A}}_{\mathrm{vib}} + a + RT'$$

$$G^0_{\mathrm{B}} = -RT \ln \; f^{\mathrm{B}}_{\mathrm{t}} f^{\mathrm{B}}_{\mathrm{r}} f^{\mathrm{B}}_{\mathrm{vib}} + b + RT$$

$$G^0_{\mathrm{C}} = -RT \ln f^{\mathrm{C}}_{\mathrm{t}} f^{\mathrm{C}}_{\mathrm{r}} f^{\mathrm{C}}_{\mathrm{vib}} + c + RT$$

But $c - a - b$ is the total heat of reaction at absolute zero, including the zero-point energies; let us call this quantity ΔE^0_0. Hence,

$$\Delta G^0 = -RT \ln \frac{f^{\mathrm{C}}_{\mathrm{t}} f^{\mathrm{C}}_{\mathrm{r}} f^{\mathrm{C}}_{\mathrm{vib}}}{f^{\mathrm{A}}_{\mathrm{t}} f^{\mathrm{A}}_{\mathrm{r}} f^{\mathrm{A}}_{\mathrm{vib}} \, f^{\mathrm{B}}_{\mathrm{t}} f^{\mathrm{B}}_{\mathrm{r}} f^{\mathrm{B}}_{\mathrm{vib}}} + \Delta E^0_0 - RT$$

Comparison of this equation with equation 9.23 gives immediately

$$K = \frac{f^{\mathrm{C}}_{\mathrm{t}} f^{\mathrm{C}}_{\mathrm{r}} f^{\mathrm{C}}_{\mathrm{vib}}}{f^{\mathrm{A}}_{\mathrm{t}} f^{\mathrm{A}}_{\mathrm{r}} f^{\mathrm{A}}_{\mathrm{vib}} \, f^{\mathrm{B}}_{\mathrm{t}} f^{\mathrm{B}}_{\mathrm{r}} f^{\mathrm{B}}_{\mathrm{vib}}} \, \mathrm{e}^{-\Delta E^0_0/RT} \, \mathrm{e}^{-\Delta n} \tag{9.26}$$

where Δn is the difference between the number of product molecules and the number of reactant molecules in the equation of reaction. E^0_0 can be estimated from the heat of reaction measured, of course, at a finite temperature by using heat capacities for A, B and C calculated from their partition functions. Fortunately, for many reactions the heat of reaction is not very sensitive to temperature and so a measured value itself can be substituted without serious error.

This method of estimating equilibrium constants is particularly useful when information is required on the efficiency of isotopic substitution reactions. Take, for example, exchange between H_2 and D_2:

$$H_2 + D_2 \rightarrow 2HD$$

In this case the total energy change for the reaction at 0 K is the algebraic sum of the zero-point energies:

$$\Delta E_0^0 = 2E_0^0(\mathrm{HD}) - E_0^0(\mathrm{H_2}) - E_0^0(\mathrm{D_2})$$

$$= \frac{hc}{2}(2\bar{\upsilon}_{\mathrm{HD}} - \bar{\upsilon}_{\mathrm{H_2}} + \bar{\upsilon}_{\mathrm{D_2}})$$

Thus

$$K = \frac{f_{\mathrm{HD}}^2}{f_{\mathrm{H_2}} f_{\mathrm{D_2}}} \, \mathrm{e}^{-hc(2\bar{\upsilon}_{\mathrm{HD}} - \bar{\upsilon}_{\mathrm{H_2}} - \bar{\upsilon}_{\mathrm{D_2}})/2kT}$$

$$= \frac{\left[V\left(\frac{2\pi M_{\mathrm{HD}} kT}{h^2}\right)^{\frac{3}{2}}\left(\frac{8\pi^2 I_{\mathrm{HD}} kT}{\sigma_{\mathrm{HD}} h}\right)(1 - \mathrm{e}^{hc\bar{\upsilon}_{\mathrm{HD}}/kT})^{-1}\right]^2 \times \mathrm{e}^{-hc(2\bar{\upsilon}_{\mathrm{HD}} - \bar{\upsilon}_{\mathrm{H_2}} - \bar{\upsilon}_{\mathrm{D_2}})/kT}}{V\left(\frac{2\pi M_{\mathrm{H_2}} kT}{h^2}\right)^{\frac{3}{2}}\left(\frac{8\pi^2 I_{\mathrm{H_2}} kT}{\sigma_{\mathrm{H_2}} h}\right)(1 - \mathrm{e}^{-hc\bar{\upsilon}_{\mathrm{H_2}}/kT})^{-1} \times \\ \times V\left(\frac{2\pi M_{\mathrm{D_2}} kT}{h^2}\right)^{\frac{3}{2}}\left(\frac{8\pi^2 I_{\mathrm{D_2}} kT}{\sigma_{\mathrm{D_2}} h^2}\right)(1 - \mathrm{e}^{-hc\bar{\upsilon}_{\mathrm{D_2}}/kT})^{-1}}$$

[As an exercise satisfy yourself that the same expression results if one assumes $\Delta E = 0$ and uses equation 1.52 for f_{vib}.] This equation simplifies, because for each vibrational partition function $hv \gg kT$ at room temperature and, hence, $f_{\mathrm{vib}} \sim 1$. Furthermore,

$$\sigma_{\mathrm{H_2}} = \sigma_{\mathrm{D_2}} = 2; \qquad \sigma_{\mathrm{HD}} = 1$$

Thus

$$K = \left(\frac{M_{\mathrm{HD}}^2}{M_{\mathrm{H_2}} M_{\mathrm{D_2}}}\right)^{\frac{3}{2}} \cdot \frac{4 I_{\mathrm{HD}}^2}{I_{\mathrm{H_2}} I_{\mathrm{D_2}}} \cdot \mathrm{e}^{-hc(2\bar{\upsilon}_{\mathrm{HD}} - \bar{\upsilon}_{\mathrm{H_2}} - \bar{\upsilon}_{\mathrm{D_2}})/2kT}$$

Now:

$$M_{HD} = 3, \qquad M_{H_2} = 2, \qquad M_{D_2} = 4$$

and

$$I_{HD} = \tfrac{2}{3} I_{D_2}, \qquad I_{H_2} = \tfrac{1}{2} I_{D_2}$$

also

$$\bar{\upsilon}_{HD} = 3817 \text{ cm}^{-1}, \qquad \bar{\upsilon}_{H_2} = 4395 \text{ cm}^{-1} \quad \text{and} \quad \bar{\upsilon}_{D_2} = 3118 \text{ cm}^{-1}$$

Hence,

$$K = \frac{9}{8} \times \frac{4 \times 8}{9} \, e^{-hc(121)/2kT}$$

$$= 2.96 \text{ at } T = 300 \text{ K}$$

Note that this value will not be accurate because we have used the integrated form of the rotational partition function, which is not a good approximation for H_2 at room temperature. *Table 9.3* shows the excellent agreement between the experimental results and accurate calculations of K obtained from equation 9.26.

Table 9.3 EQUILIBRIUM CONSTANTS*: $H_2 + D_2 \rightarrow 2HD$

	K(expt.)	K(calc.)
195	2.92 ± 0.08	2.84
273	3.24 ± 0.08	3.18
298	3.28 ± 0.08	3.26
383	3.50 ± 0.06	3.46
543	3.85 ± 0.1	3.67

* *J. Chem. Phys.*, **2**, 48, 362 (1934).

Entropies

Entropy is a quantity of central importance in chemistry (and in such fields as aerodynamics, where entropies at elevated temperatures are needed). The contribution of each degree of freedom to the total entropy may be simply calculated from the relevant partition functions via equation 9.7 or equation 1.34, which we can rewrite in the form

$$S = kT\left(\frac{\partial \ln (f^N)}{\partial T}\right)_V + k \ln (f^N)$$

For example, if we substitute in the equation the translational partition function, we can calculate the translational entropy for one mole of an ideal gas.

In equations 9.7 and 1.34 we have assumed that the partition function f_N for N molecules is related to the molecular partition function f by the equation $f_N = f^N$. However, this is only true if the particles being studied are distinguishable. In the case of molecules in a gas, it is impossible to distinguish one molecule from another, since no molecule has a particular energy or a particular position. For N molecules there are $N!$ indistinguishable arrangements in the gas. Hence, for a gas containing N *identical* atoms or molecules,

$$f_N = \frac{f^N}{N!}$$

and

$$S_t = kT\left(\frac{\partial}{\partial T}\ln\left(\frac{f_t^N}{N!}\right)\right)_V + k\ln\left(\frac{f_t^N}{N!}\right)$$

$$= kT\left(\frac{\partial}{\partial T}\ln\frac{V}{N!}\left(\frac{2\pi mkT}{h^2}\right)^{3N/2}\right)_V + k\ln V\left(\frac{2\pi mkT}{h^2}\right)^{3N/2} - k\ln N!$$

therefore

$$S_t = kT\left(\frac{3N}{2}\cdot\frac{1}{T}\right) + Nk\ln V\left(\frac{2\pi mkT}{h^2}\right)^{3} - k(N\ln N - N)$$

$$= \tfrac{3}{2}Nk + Nk\ln\frac{V}{N}\left(\frac{2\pi mkT}{h^2}\right)^{3} - Nk$$

Hence, for one mole of gas

$$S_t = \tfrac{5}{2}R + R\ln\frac{V}{N}\left(\frac{2\pi mkT}{h^2}\right)^{\frac{3}{2}}$$

This is the well-known Sackur–Tetrode equation. For a monatomic gas it gives the total entropy of the gas.

The extension of these ideas to diatomic molecules is straightforward, and *Table 9.4* makes a comparison between the entropies calculated from equation 1.34 and those determined calorimetrically.

Calorimetric determination of the entropy is usually achieved by measuring the molar heat capacity of the substance from 10 K up to the required temperature, measuring the heats of fusion and vaporisation, and the heats of any phase transitions in the solid. The integral from 0 K to T K of the quantity dQ/T then has to be evaluated: the

Table 9.4 MOLAR ENTROPIES OF DIATOMIC GASES AT 298 K AND 1 atm in cal mol^{-1} deg^{-1}

Molecule	*Spectroscopic*	*Calorimetric*
N_2	45.79	45.9
O_2	49.03	49.1
HCl	44.64	44.5
HBr	47.48	47.6
HI	49.35	49.5

experimental heat capacity or heat of transition is used from 10 K upwards, and the theoretical heat capacity calculated according to Debye's theory is used from 0 K to 10 K.

It can be seen in *Table 9.4* that there is excellent agreement for these molecules between the spectroscopic and calorimetric entropies; note that the latter quality is derived assuming that at absolute zero the entropy is zero, i.e. the assumption of the third law of thermodynamics. There are, however, a number of molecules for which the calorimetric and spectroscopic entropies differ by an amount $nR \ln x$, where n and x are simple integers or fractions. This has not been fully understood, but it seems likely from recent work by Morrison in Ottawa that many of these discrepancies may be due to unsuspected transitions which take place in the solid at about 1 K—this has been proved for the molecule CH_3D and may be true for many other molecules. This remark should be borne in mind when entropies rather than heat capacities (cf. Chapter 8) are used to deduce information about the nature of spectroscopically inactive vibrations.

Having established the correctness of the spectroscopic method, we are now in a position to apply it at any temperature for which the formulae are valid—contrast the calorimetric method, which makes it extremely difficult to measure heat capacities more than a few hundred degrees above room temperature. Thus, the major thermodynamic compilations list entropies and other thermodynamic data up to 6000 K, sometimes up to 25 000 K. A particularly elegant example of the use of these computed data is in the determination of the heats of chemical reactions from equilibrium constants. The example discussed here is the equilibrium

$$CH_4 + I_2 \rightleftharpoons CH_3I + HI$$

for which equilibrium constants K_T can be determined experimentally in the temperature range $T = 575\text{–}750$ K. One of the quantities regularly tabulated for each molecule is $-(G_T^0 - H_{298}^0)/T$, and if Δ is used to denote the difference between reactants and products for any individual thermodynamic quantity, it can easily be shown that

$$\Delta H_{298}^0 = -RT \ln K + T\Delta[-(F_T^0 - H_{298}^0)/T]$$

Table 9.5 lists the observed equilibrium constants, which vary with temperature by a factor of 10 over this range, together with the

Table 9.5 THE EQUILIBRIUM $CH_4 + I_2 \rightleftharpoons CH_3I + HI$

T, K	$\ln K_T$	$\Delta[-(F_T^0 - H_{298}^0)/T]$ cal mol^{-1} deg^{-1}	ΔH_{298}^0, cal mol^{-1}
748	−2.945 5	3.480	12 685
732	−3.026 7	3.468	12 677
717	−3.087 1	3.457	12 607
708	−3.158 8	3.451	12 677
688	−3.259 5	3.436	12 626
620	−3.730 0	3.387	12 682
616	−3.711 0	3.385	12 545
607	−3.806 3	3.378	12 623
585	−3.968 8	3.364	12 592
		Av.	12 635

spectroscopically calculated free-energy function and the derived values of ΔH_{298}^0: the constancy of the latter is remarkable. Two points can be made here. First, that this is a better way of obtaining heats of reaction from equilibrium constants than the use of the van't Hoff relation, since this method relates ΔH directly to $\ln K_T$, whereas the van't Hoff method relates ΔH to differences between $\ln K_T$ at various temperatures. Second, it is clear that if we know ΔH_{298}^0 for any given reaction, we can turn the calculation round, and deduce the composition of the equilibrium mixture at any temperature we like—hence the importance of such methods in calculating the composition of hot reaction mixtures such as rocket exhaust gases.

Exercises

1. Assuming the form of the appropriate partition function, show that the entropy of an ideal monatomic gas at temperature T and pressure P is:

$$S = \tfrac{5}{2}R \ln T - R \ln P + \tfrac{3}{2}R \ln M - 1.165R$$

A monatomic gas has an average atomic weight M and is composed of two isotopes having atomic weights $(M+1)$ and $(M-1)$. Show that if the entropy is calculated (a) on the basis of two isotopes, and (b) assuming the gas to be composed entirely of atoms having the average atomic weight, the two estimates will differ by $R \ln [2(1 - M^{-1})^{\frac{3}{4}}]$.

2. Calculate the equilibrium constant at 300 K for the reaction

$$I_2 \rightleftharpoons 2I$$

given that for I_2 the fundamental frequency of vibration is 214 cm^{-1}, the moment of inertia is 748×10^{-40} g cm^2 and $D_0 = 35.0$ kcal mol^{-1}.

3. The moment of inertia of $H^{35}Cl$ is 2.71×10^{-40} g cm^2 and its fundamental vibrational frequency is 2886 cm^{-1}. Calculate the entropy and heat capacity at constant volume for HCl as an ideal gas at 1 atm pressure and 27°C.

4. The Cl atom is in the level $^2P_{\frac{3}{2}}$ in its ground state. The next level ($^2P_{\frac{1}{2}}$) lies 2.5 kcal g-atom^{-1} higher. The third level is separated by a much larger energy difference. Calculate the electronic contribution to the heat capacity of the chlorine atom at 300, 500, 1000 and 1500 K.

BIBLIOGRAPHY

The following books may be found useful. They range from easy to very difficult. No one of them is relevant to the whole text, and some of them are recommended because they treat one or two topics well.

Statistical Mechanics

Davidson, N., *Statistical Mechanics,* McGraw-Hill, New York, 1962

Dole, M., *Statistical Thermodynamics,* Prentice-Hall, Englewood Cliffs, N.J., 1954

Fowler, R. H. and Guggenheim, E. A., *Statistical Thermodynamics,* Cambridge University Press, 1939

Gurney, R. W., *Introduction to Statistical Mechanics,* McGraw-Hill, New York, 1949

Hanna, M. W., *Quantum Mechanics in Chemistry,* Benjamin, New York, 1965

Hinshelwood, C. N., *Structure of Physical Chemistry,* 1951, Clarendon Press, Oxford

Margenau, H. and Lindsay, R. B., *Foundations of Physics,* Wiley, New York, 1936

Nash, L. K., *Elements of Statistical Thermodynamics,* Addison-Wesley, London, 1968

Pitzer, K. S., *Quantum Chemistry,* Prentice-Hall, Englewood Cliffs, N.J., 1953

Rushbrook, G. S., *Introduction to Statistical Mechanics,* Clarendon Press, Oxford, 1949

Slater, J. C., *Introduction to Chemical Physics,* McGraw-Hill, New York, 1939

Spectroscopy

Barrow, G. M., *Introduction to Molecular Spectroscopy,* McGraw-Hill, New York, 1962

Condon, E. U. and Shortley, G. H., *Theory of Atomic Spectra,* Cambridge University Press, 1935

Dixon, R. N., *Spectroscopy and Structure,* Methuen, London, 1965

Dodd, R. E., *Chemical Spectroscopy,* Elsevier, New York, 1962

Dunford, H. B., *Elements of Diatomic Molecular Spectra,* Addison-Wesley, London, 1968

Gordy, W., Smith, W. V. and Trambarulo, R. F., *Microwave Spectroscopy,* Wiley, New York, 1953

Herzberg, G., *Atomic Spectra and Atomic Structure,* Dover, New York, 1944

Herzberg, G., *Infrared and Raman Spectra,* Von Nostrand, New York, 1945

Herzberg, G., *Spectra of Diatomic Molecules,* Van Nostrand, New York, 1959

Townes, C. H. and Schawlow, A. L., *Microwave Spectroscopy,* McGraw-Hill, New York, 1955

SYMBOLS

Symbol	*Quantity*
a	constant, percentage of light transmitted, linear dimension, coordinate axis
a	angular (subscript)
B	rotational constant
b	linear dimension, coordinate axis
c	velocity of light, linear dimension, coordinate axis
c_p	heat capacity at constant pressure
c_V	heat capacity at constant volume
D	optical density, dissociation energy, centrifugal distortion constant
d	electron level
E	energy (total), static electric field
e	base of natural logarithms, equilibrium (subscript), electronic energy (subscript)
F	Helmholtz free energy
f	partition function
f	electron level
G	Gibb's free energy
g	degeneracy
g	gerade (subscript)
H	enthalpy
h	Planck's constant, height
I	intensity of light, moment of inertia
$I_{\parallel}$	intensity of a Raman band in a plane parallel to polarisation
$I_{\perp}$	intensity of a Raman band in a plane perpendicular to polarisation
i	positive integer (subscript), quantum number
i	incident (subscript)

Symbol	*Quantity*
int	internal (subscript)
J	rotational quantum number, total angular momentum quantum number
j	positive integer (subscript)
K	equilibrium constant
k	constant, Boltzmann constant, force constant; positive integer
L	orbital angular momentum quantum number
l	orbital angular momentum quantum number, positive integer (subscript)
M	magnetic quantum number
m	mass, magnetic quantum number
max	maximum (subscript)
N	number (of systems)
n	number, principal quantum number
obs	observed (subscript)
p	pressure
p	electron level
R	universal gas constant
r	distance separating atomic particles, radius
r	rotation (subscript)
S	entropy, spin angular momentum quantum number
s	spin angular momentum quantum number
s	electron level, scattered (subscript)
T	temperature
t	time
t	translational (subscript)
u	ungerade (subscript)
V	volume, potential energy
ν §	quantum number of a vibrational state
x	coordinate axis, algebraic unknown quantity
y	coordinate axis, algebraic unknown quantity
z	coordinate axis
α	constant, polarisability of a molecule
β	constant
Δ	molecular spectroscopic state
$\Delta\upsilon$*	convergence limit
ϵ	energy
θ	characteristic temperature
Λ	orbital angular momentum quantum number of a molecule

§ Denoted by ν in the figures.

Symbol	*Quantity*
λ	wavelength
μ	dipole moment, reduced mass
υ† §	frequency
$\bar{\upsilon}$† §	frequency (in wave-number units)
Π	molecular spectroscopic state
ρ	depolarisation ratio, density
Σ	molecular spectroscopic state
σ	symmetry number
ψ	wave function
Ω	number of significant permutations of a system, number of equally probable arrangements of a system
ω†	frequency (in hertz)
$'$	upper state
$''$	lower state

† Both ω and $\bar{\upsilon}$ are used as symbols for frequency. ω, in hertz, is the fundamental quantity; $\bar{\upsilon}$, in cm^{-1}, derives from experimental practice, because many spectrometers are provided with linear scales in cm^{-1}. The relationship between ω and $\bar{\upsilon}$ is: $\omega = c\bar{\upsilon}$.

§ Denoted by ν and $\bar{\nu}$ in the figures.